Analytical Chemistry and Microchemistry

nova science publishers

www.novapublishers.com

Analytical Chemistry and Microchemistry

Chromatographic Methods and Research
Raven W. Obrien (Editor)
2023. ISBN: 979-8-89113-193-4 (Hardcover)
2023. ISBN: 979-8-89113-203-0 (eBook)

Airborne and Biological Monitoring to Assess Occupational Exposure to Isocyanates
Jimmy Hu, PhD
2023. ISBN: 979-8-89113-010-4 (Softcover)
2023. ISBN: 979-8-89113-099-9 (eBook)

Chemometrics: Advances in Applications and Research
Larry D. Crenshaw (Editor)
2022. ISBN: 979-8-88697-274-0 (Hardcover)
2022. ISBN: 979-8-88697-311-2 (eBook)

Functional Analytical Techniques in Pharmaceutical Chemistry
Malik Saadullah (Editor)
2022. ISBN: 979-8-88697-059-3 (Hardcover)
2022. ISBN: 979-8-88697-288-7 (eBook)

Modern Approaches in Fluid Chromatography: Impact and Applications
Rohit Bhatia, PhD and Bhupinder Kumar, PhD
2023. ISBN: 979-8-88697-856-8 (Hardcover)
2023. ISBN: 979-8-89113-022-7 (eBook)

A Guide to Near-Infrared Spectroscopy
Juan Francisco García Martín, PhD (Editor)
2023. ISBN: 979-8-88697-775-2 (Hardcover)
2023. ISBN: 979-8-88697-968-8 (eBook)

More information about this series can be found at
https://novapublishers.com/product-category/series/analytical-chemistry-and-microchemistry/

Raven W. Obrien
Editor

Chromatographic Methods and Research

nova
science publishers
www.novapublishers.com

Library of Congress Cataloging-in-Publication Data

ISBN: 979-8-89113-193-4

Published by Nova Science Publishers, Inc. † New York

Contents

Preface

This book contains six chapters that detail chromatographic methods and research. Chapter One discusses how liquid chromatography-ultraviolet (LC-UV) bioanalytical methods can be considered an important tool for beta-lactams serum measurements to evaluate in a real time drug effectiveness in septic burn patient's bedside. Chapter Two focuses on the preconcentration strategies coupled with chromatographic methods used for the detection of various types of organic compounds. Chapter Three details the bioanalytical methods, based on liquid chromatography, for the therapeutic drug monitoring of combined therapies of antibiotics in pediatric burns. Chapter Four presents liquid chromatography-tandem mass spectrometry. Chapter Five briefs about the determination of an antiviral drug, favipiravir (FVPR) in pure and pharmaceutical dosage forms and different biological matrix of samples, by employing various chromatographic techniques. Lastly, Chapter Six outlines key considerations for the analysis of the hormone α-MSH in brain tissue by the liquid chromatography tandem mass spectrometry (LC-MS/MS) method.

Chapter 1

Beta-Lactam Agents Serum Measurements by Liquid Chromatography to Evaluate Meropenem and Piperacillin Effectiveness in Septic Burn Adults with a Pharmacokinetic-Pharmacodynamic Approach

Leonard de Vinci Kanda Kupa[1], PhD
Vanessa Kasubeck de Souza[1], PhMs
Thais Vieira Camargo[1], PhMs
Claudia Garcia Messiano[1], PhMs
Ronaldo Morales Junior[1], PhD
Valentina Porta[1], PhD
Mauricio Yonamine[2], PhD
Beatriz Aparecida Passos Bismara Paranhos[2], PhD
Aline Sandré Gomides Abad[3], BS
Gabriela Aparecida Ferreira[3], BS
Edvaldo Vieira de Campos[3], MD
Élson Mendes da Silva Junior[3], MD
João Manoel da Silva Junior[3], MD
Debora Cristina Sanches Pinto[3], MD
David de Souza Gomez[3,*], MD

* Corresponding Author's Email: davgomez@usp.br.

In: Chromatographic Methods and Research
Editor: Raven W. Obrien
ISBN: 979-8-89113-193-4

and Silvia Regina Cavani Jorge Santos[1,†], PhD

[1]Clinical Pharmacokinetics Center, School of Pharmaceutical Sciences, University of Sao Paulo, Sao Paulo, Brazil

[2]Department of Clinical-Toxicology Analysis, Multi-User Laboratory of Analytical Center, School of Pharmaceutical Sciences, University of Sao Paulo, Sao Paulo, Brazil

[3]Plastic Surgery and Burns Division of Clinics Hospital, Surgery Department of Medical School, University of Sao Paulo, Sao Paulo, Brazil

Abstract

Background: Meropenem and Piperacillin-Tazobactam are largely prescribed for septic patients undergoing intensive care therapy of septic shock, and therapeutic drug monitoring (TDM) permits to evaluate drug serum levels and effectiveness against Gram-negative strains isolated from cultures. Subject of study was to evaluate drug effectiveness by liquid chromatography-ultraviolet detection (LC-UV) for piperacillin (PIP) and meropenem (MER) based in serum measurements. *Bioanalytical methods:* Due to its good sensitivity and selectivity, liquid chromatography (LC) technique is considered the gold standard for TDM of beta-lactam agents. Then, it was developed a new and quite simple bioanalytical method to quantify PIP in serum by LC-UV detector set at 210 nm. MER serum measurements were done by LC-UV detector set at 307 nm by applying a re-validated bioanalytical method to evaluate effectiveness of beta-lactams agents in septic major burns. Validation study was based on international standards considering the parameters as linearity, lower limit of quantification, lower limit of detection, intra-day/inter days precisions, intra-day/inter days systematic error, stability of standards on the bench (4 hrs.), short serum samples stability on the bench (6 hrs.), post-process stability of serum extracts in the rack of autosampler up to 24 hrs., serum samples long stability (ultralow freezer storage), cycles freezing/de-freezing stability, and the robustness of chromatographic system. *Patients:* Forty-five major burn patients of both genders, with renal function preserved or augmented by vasopressors, and infections caused by Gram-negative nosocomial pathogens were included. Patients undergoing therapy at the earlier stage of septic shock were distributed in two groups based on meropenem or piperacillin prescription. Twenty-two patients undergoing the therapy of septic shock with meropenem 1 g q8h, or piperacillin-tazobactam 4.5 g q6h by intermittent 0.5 hr.-infusion were included in the group 1 (G1, 22) as recommend in hospital. Additionally, another twenty-three patients

† Corresponding Author's Email: pharther@usp.

received the same dose regimens of both antimicrobial agents by an extended 3 hrs.-infusion, group 2 (G2, n = 23) patients. Blood was sampling from patients of both groups, 48 hrs. after the therapy started for drug serum measurements by LC-UV at the 3rd hr. of started infusion, and one hour before the next infusion. Pharmacokinetic-pharmacodynamics approach-based on serum levels was done to attain therapeutic target of 100%$f\Delta T$ > MIC recommended for both beta-lactam agents under investigation. Then, the prediction index of antimicrobial effectiveness (%$f\Delta T$ > MIC) was applied to evaluate target attainment. *Results:* Coverage occurred for all patients of both groups after intermittent 0.5 hr.-infusion up to MIC 2 mg/L (minimum inhibitory concentration), while the clinical cure occurred for patients after the extended 3 hrs.- infusion with eradication of Gram-negatives pathogens up to MIC 4 mg/L by the meropenem, and up to MIC 16 mg/L by the piperacillin according to Clinical Standards Laboratory Institute (CSLI, database) considered in the Microbiology of our hospital. Superiority of 3 hrs.-infusion was registered by comparison with the intermittent 0.5 hr.-infusion recommended yet in hospitals. *Conclusion:* LC-UV bioanalytical methods can be considered an important tool for beta-lactams serum measurements to evaluate in a real time drug effectiveness in septic burn patient's bedside.

Keywords: TDM beta-lactams by LC-UV, Drug effectiveness related to time of infusion, PK/PD approach based on serum levels, therapy guided by PK/PD tools and cultures

Abbreviations

GSA	Global Sepsis Alliance
ICU	Intensive Care Unit
MDR	Multidrug Resistance
MIC	Minimum Inhibitory Concentration
MV	Mechanical Ventilation
PD	Pharmacodynamics
PK	Pharmacokinetics
PTA	Probability of Target Attainment
SAPS3	Simplified Acute Physiology Score 3
SARS-CoV-2	Severe Acute Respiratory Syndrome Coronavirus 2
SIRS	Systemic Inflammatory Response Syndrome
TBSA	Total Burn Surface Area

TDM	Therapeutic Drug Monitoring
WHO	World Health Organization

Introduction

Septic shock is potentially fatal organ dysfunction caused by a dysregulated host response to infection. Clinical outcome in most high-risk cases is the mortality of patients including elderly and pediatric with nosocomial bacterial infections associated with viral infections as influenza and most recently SARS-CoV-2 related to One Global Health Treat: Sepsis, Pandemics, and Antimicrobial Resistance discussed during 48 hours at the 4th World Sepsis Congress occurred in Berlin last April 25 and 26, 2023 organized by Global Sepsis Alliance (2023). In view of the growing challenge to the prescription of antimicrobials for adequate treatment and effective control of bacterial infectious conditions during COVID-19 pandemic, International Guidelines for Management of Sepsis and Septic Shock (2021), Global Sepsis Alliance (2020), and World Health Organization (2023) have reinforced that combating bacterial resistance, and the prevention of the development of multidrug-resistant strains (MDR) is urgent (Abdulla et al., 2020; Evans et al., 2021). It is well known that at the last twenty-five years, it was reported a significant increase in the minimum inhibitory concentration (MIC). In a monitoring program of antimicrobial therapy for several infections based on serum levels of beta-lactams and vancomycin in ICU patients. It was reported for vancomycin that almost 90% of patients did not reach the therapeutic target against susceptible strains of Gram-positive, MIC 2 mg/L. In addition, also for beta-lactam agents as meropenem or piperacillin-tazobactam largely prescribed at the recommended doses by intermittent 0.5 hr.-infusion against Gram-negative bacteria in a combined therapy with vancomycin for the treatment of septic shock caused by nosocomial pathogens, it is well known that the coverage occurs up to MIC 2 mg/L strains by the meropenem and by the piperacillin. Information reinforces that antimicrobial serum levels monitoring is essential to assess pharmacokinetics changes that occurs during the septic shock with impact on coverage of the prescribed antimicrobial agent expressed through target attainment reached by pharmacokinetic-pharmacodynamics (PK/PD) approach. Therefore, new therapeutic strategies have been proposed for the most prescribed antimicrobial agents as meropenem and piperacillin-tazobactam related to the dose regimen and the duration of infusion (Abdul-Aziz et al., 2016; Barlam et al., 2016; Cain et al.,

2015; Carlier et al., 2015; Roberts et al., 2010; Santos et al., 2023). Some prospective controlled studies including therapeutic drug monitoring (TDM) were considered to compare drug efficacy based on serum levels of several antimicrobial agents usually prescribed in the therapy of septic shock. Considering that therapeutic meropenem serum monitoring after 3 hrs.-extended infusion will be a strategy that ensures an adequate serum level against intermediate susceptible strains with coverage guaranteed up to MIC 4 mg/L in the combat against bacteria of intermediate susceptibility, mainly related to *K. pneumoniae* and *P. aeruginosa* in major septic burns with renal function preserved or augmented by vasopressors (Fournier et al., 2015; Santos et al., 2022). In addition, to allowing the evaluation of antimicrobial effectiveness considering the dose regimen prescribed to septic ICU patients, drug serum levels are a laboratory strategy of great value in the individualization of therapy in a real time, guaranteeing the expected clinical outcome, and to combating the development of bacterial resistance, also reducing the duration of antimicrobial therapy, and consequently hospital costs (Abdulla et al., 2020; Barlam et al., 2016). Impact of the introduction of real-time therapeutic drug monitoring on empirical doses of carbapenems in septic burn patients was reported to adequate empirical antibiotic dose selection for these patients, once is difficult due to extreme variability in drug pharmacokinetics. Therapeutic drug monitoring (TDM) by liquid chromatography may aid antibiotic prescription and implementation of initial empirical antimicrobial dosage recommendations. In several studies it was evaluated how gradual TDM introduction altered empirical dosages of carbapenem agents as meropenem or even piperacillin-tazobactam, both largely prescribed to treat infections caused by nosocomial pathogens in critically ill major burn patients of the ICU Burn Center of hospital (Fournier et al., 2015). Several studies related to meropenem, and piperacillin serum levels by TDM were related to liquid chromatography - ultraviolet detection (LC-UV), or even by liquid chromatography - diode array detection or mass spectrometer detection, LC-MS/MS (Caro et al., 2020, Casals et al., 2014; Chung et al., 2015; De Waele et al., 2014; Ferrone et al., 2017; Fournier et al., 2015; Gonçalves-Pereira et al., 2014; Jaruratanasirikul & Sriwiriyajan, 2003; McWhinney et al., 2010; Nilsson et al., 1991; Occhipinti et al., 1997; Roth et al., 2017; Santos et al., 2011; Santos et al., 2022; Santos et al., 2023, Schoenenberger-Arnaiz et al., 2019; Sime et al., 2015; Souza et al., 2019, Taccone et al., 2010; Zou et al., 2019). In addition, the last technique LC-MS/MS was applied by many authors for beta-lactams serum measurements as described in a recent review study published by Caro et al., (2020). On the other hand, this technique is more complex and expensive compared to conventional LC-UV. Consequently, the initial investment will be high, but it would be done a routine of individualization of antimicrobial therapy by applying the LC-MS/MS technique. However, this should be done in large tertiary hospitals with a great number of ICUs to care for a huge number of septic ICU patients treated daily for nosocomial infections (Caro et al., 2020). Finally, it is noteworthy that, so far, serum levels of these agents are not routinely monitored in hospitals for these patients admitted to ICUs (Abdul-Aziz et al., 2016; Abdulla et al., 2020; Carlier et al., 2015). Clinical management for critically ill patients in intensive care has been guided by cultures, through the isolation of the agent, and susceptibility test of the pathogen to the antimicrobial, supported also by ICU routine of serum biomarkers of the systemic inflammatory response syndrome (SIRS). Then, if culture results and coverage by PK/PD tools based on beta-lactam serum levels will be done in parallel, it is possible to guide properly the antimicrobial therapy to reach soon the desired clinical outcome. If serum levels of the prescribed antimicrobial agent are equal to or greater than those required to eradicate the susceptible pathogens, microbiological cure occurs, and the desired clinical outcome will be achieved by appropriately treating the septic shock and healing the patient. On the other hand, for serum levels below the recommended level, therapeutic failure will occur contributing to development of microbial resistance and bacterial emergency (Abdul-Aziz et al., 2015; Abdulla et al., 2020). Then, selective techniques as liquid chromatography must be chosen as laboratory strategy in hospitals of high complexity to facilitate therapeutic drug serum monitoring done in a real time for ICU critically ill septic patients with nosocomial infections.

Rational of Study

Subject of study was to evaluate drug effectiveness by application of pharmacokinetic-pharmacodynamics (PK/PD) tools based on serum levels done by liquid chromatography-ultraviolet detector (LC-UV) using bioanalytical methods developed for meropenem and piperacillin serum measurements once liquid chromatography technique is considered the gold standard for TDM of antimicrobial agents due to its sensitivity and selectivity.

Bioanalytical Method for Beta-Lactam Serum Levels

Meropenem

To each 200 μL of serum sample, 25 μL (50 μg/mL) of cefepime (internal standard, US Pharmacopeia) was added to an Eppendorf tube followed by 600 μL of acetonitrile and centrifuged 8,000 rpm (5°C) for 15 min. Supernatant was evaporated to dryness in a water bath (45°C) under a stream of nitrogen. A volume of 600 μL of 10% of acetonitrile in water was added to dissolve the residue, and 5 μL of serum extracts were injected into a Shimpack ODS column (150 mm x 6 mm ID, 5 micron). Elution of peaks was done by an isocratic system with a mobile phase of a mixture of acetate buffer 0.01 M, pH 5.0 and acetonitrile (9:1, v/v) at a flow rate of 0.8 mL/min. Peaks were monitored by UV detection at 307 nm for this carbapenem agent. Validation study was based on international standards considering the following parameters as linearity, lower limits of quantification and of detection, intra-day/inter days precisions, intra-day/inter days systematic error, stability of standards including the internal standard on the bench during 4 hrs., short samples stability on the bench (6 hrs.), post-process stability of serum extracts in the rack of autosampler rack (24 hrs.), samples long stability storage in an ultralow freezer, cycles freezing/de-freezing stability, and the robustness of chromatographic system.

Piperacillin

To each 200 μL of serum sample, 40 μL (50 μg/mL) of oxacillin (internal standard, US Pharmacopeia) was added to an Eppendorf tube followed by 600 μL of acetonitrile and centrifuged 8,000 rpm (5°C) for 15 min. Supernatant 600 μL was transferred to a mini vial (1 mL) and 5 μL of extracts were injected automatically into a Shimpack CN column (150 mm x 6 mm ID, 5 micron) overnight. Piperacillin was eluted with dihydrogen-phosphate buffer 0.01 M and acetonitrile (68:32, v/v) mobile phase final pH 3.5, at a flow rate of 1.0 mL/min. Peaks were monitored by UV detection set at 210 nm. Validation study was based on international standards. LC-UV210 method was successfully developed and validated with a run time lower than 10 minutes per sample analyzed in a rack of 100 mini vials overnight. Validation study was based on international standards considering the following parameters as linearity, lower limits of quantification and of detection, intra-day/inter days precisions, intra-day/inter days systematic error, stability of standards including the internal standard on the bench during 4 hrs., short samples stability on the bench (6 hrs.), post-process stability of serum extracts in the rack of autosampler rack (24 hrs.), samples long stability storage in an ultralow freezer, cycles freezing/de-freezing stability, and the robustness of chromatographic system.

Clinical Trial: Patient Eligibility, Beta-Lactams Therapy, Blood Sampling, Pharmacokinetics, PK/PD Approach

The clinical protocol was a prospective, open-label study. Ethical approval register CAAE 07525118.3.0000.0068, Brazilian Platform was obtained by approval of the Ethical Committee of Hospital of Clinics, Medical School, University of Sao Paulo; no conflicts of interest to declare were obtained from all authors. The study was conducted from June 2018 to March 2019, and informed written consent was obtained from all legally designated patient representatives. Adult patients from the Intensive Care Burn Unit, presenting severe injury by fire or electricity, and a sepsis diagnosis as reported in the "American Burn Association consensus conference to define sepsis and infection in burns" in clinical evaluation and laboratorial data were eligible for inclusion (Greenhalgh et al., 2007). Patients with meropenem or piperacillin intolerance or renal impairment were excluded. The study was based on the recommended antimicrobial treatment to suspected or documented Gram-positive and Gram-negative nosocomial infections of hospital. Meropenem or piperacillin-tazobactam therapy were prescribed at dose regimens recommended in our hospital for critically ill burn patients with renal function preserved or augmented by vasopressors requirements, that occurs mainly at the earlier stage of septic shock. Forty-five ICU major burn patients of both genders, at the earlier stage of septic shock, were included in the study to investigate antimicrobial effectiveness of therapy with meropenem, 1 g q8h, or piperacillin -tazobactam, 4.5g q6h. Both therapies were guided by cultures, and by the PK/PD approach based on drug serum levels analyzed by liquid ultraviolet detection (LC-UV). Patients were allocated in two groups based on the duration of infusion for both antimicrobial agents investigated. Twenty-two septic burn patients received by intermittent 0.5 hr.-infusion: meropenem 1g q8h (n = 12), or piperacillin-tazobactam 4.5 g q6h, as recommend in hospital, group 1, n = 10 patients. Another twenty-three patients received the same dose regimens of both antimicrobial agents by an extended 3 hrs.-infusion, group 2, n = 23 patients, applying PK/PD tools and the new strategy related to drug infusion recommended (Abdul-Aziz et al., 2015). Blood was sampling 48 hrs. after the therapy started for drug serum measurements by LC-

UV at the 3rd hr. of started infusion, and one hour before the next infusion. Complete medical histories, physical examinations were obtained for each enrolled patient; laboratory data included documented strains in blood, bone, pulmonary lavage, and urine cultures; susceptibility testing done to obtain the minimum inhibitory concentration for each antimicrobial agent against each pathogen isolated. Serum samples was analyzed by liquid chromatography with detection set at 307 nm for meropenem, or detection set at 210 nm for piperacillin serum measurements. One compartment open model was chosen to investigate pharmacokinetics, and noncompartmental data analysis was applied. Pharmacokinetic-pharmacodynamics approach-based on serum levels to attain therapeutic target of 100%$f\Delta$T > MIC for piperacillin and meropenem, recommended and previously reported (Abdul-Aziz et al., 2016). Then, prediction index of antimicrobial effectiveness (%$f\Delta$T > MIC) was applied to evaluate target attainment. Drug effectiveness was evaluated by pharmacokinetic-pharmacodynamics approach, guided by PK/PD target attainment and microbiology of Gram-negative isolates from cultures.

Analytical Procedures for Drug Serum Measurements

Confidence limits obtained for meropenem and piperacillin serum measurements in the validation study are described in Table 1 including a linear response of 0.5/0.4 up to 200 mg/L (r^2:0.9998/0.9994) that was obtained with a lower limit of quantification (LLOQ) of 0.5/0.4 mg/L, and a limit of detection 0.25/0.20 mg/L, respectively for meropenem/piperacillin. Intraday, and inter days precisions were acceptable at low, medium, and high concentrations based on internal controls; the relative recovery ranging 95-105% was obtained. It was evidenced acceptable stability related to standards on the bench (4 hrs.), samples short stability (6 hrs.) on the bench, post-process stability 24 hrs. in the autosampler, samples long stability by storing in the ultralow freezer, and cycles freezing/de-freezing stability. Robustness of each chromatographic system was also included. In addition, meropenem confidence limits of the bioanalytical method are detailed in Table 1.

Table 1. Bioanalytical method for drug serum measurements by liquid chromatography LC UV-VIS

Bioanalytical Method (serum matrices)	LC UV 307 nm	LC UV 210 nm
Validation parameters	Meropenem	Piperacillin
Linearity (lower/upper limits)	0.5-200 mg/L	0.4-200 mg/L
Linear regression coefficient	0.9998	0.9994
Lower limit f quantification (LOOQ)	0.5 mg/L	0.4 mg/L
Limit of detection (LOD)	0.25 mg/L	0.2 mg/L
Relative recovery (mean, CI95%)	98(94-103)	97(95-105)
Intraday precision (CV%) n = 18	6.1(5.5-6.8)	6.6(5.9-7.2)
Inter-day precision (CV%) n = 27	7.8(6.5- 9.3)	7.1(6.0- 8.9)
Systematic error intra-day (%) n = 18	7.8(6.9-8.5)	5.8(4.9-7.6)
Systematic error inter-day (%) n = 27	8.8(7.9-9.8)	6.7(6.1-8.7)
Standard stability, bench 4 hrs. (%) n = 9	2.8(1.9-4.5)	3.4(2.6-5.2)
Internal standard stability, bench 4 hrs. (%) n = 6	3.5(3.1-4.5)	2.5(2.1-4.4)
Samples short stability, bench 6 hrs. (%) n = 9	4.5(2.8-4.2)	3.9(3.3-5.0)
Post-process stability, rack 24 hrs. (%) n = 9	5.8(4.2-6.5)	4.8(4.0-7.2)
Samples long stability, UL freezer (%) n = 118	7.5(7.2-9.4)	6.9(6.2-8.8)
Cycles freezing/de-freezing stability n = 27	5.8(5.5-6.7)	4.9(4.5-5.9)
Robustness (chromatographic system) n = 72	8.1(6.2-8.7)	7.2(6.9-8.9)

Abbreviations: CV: coefficient of variation; LOOQ: lower limit of quantification; LOD: Limit of detection; CI: Confidence interval; UL: ultralow.

Statistics: GraphPad Prism software, version 9.1.4.

Method was revalidated by liquid chromatography technique that was developed for meropenem serum measurements included a linear response of 0.5 up to 200 mg/L (r^2:0.9998) that was obtained with a lower limit of quantification (LLOQ) of 0.5 mg/L, and a limit of detection 0.25 mg/L. Intraday, and inter days precisions were acceptable at low, medium, and high concentrations based on internal controls, and the relative recovery ranging 94-103% was registered. It was evidenced acceptable stability studies related to standards, serum samples, post-process stability in the autosampler, samples long stability (ultralow freezer), cycles freezing/de-freezing stability and robustness of each chromatographic system was also included, Table 1. Both methods were successfully developed and validated with a run time lower than 10 min. per sample analyzed in a rack of 100 mini vials overnight. It was considered a coefficient of determination (r2 > 0.999) for both drug assays over the standard curve concentrations based on eight serum calibrators plus the zero (C0) were acceptable (linearity C1-C8: up to 200 mg/L), C1:0.4 mg/L (piperacillin) or C1: 0.5 mg/L (meropenem); internal controls, high, medium, and low were included. Calibration daily curve was accepted based on systematic error lower than 10% for the internal controls (high, medium, low serum levels). Drug serum levels in patients' samples were based on the accepted daily curve. Primary standard US Pharmacopoeia of meropenem, piperacillin, cefepime, oxacillin was acquired to carry out the project.

Clinical Studies

Demographic, clinical characteristics of patients at admission and during drug serum monitoring of them undergoing therapy of infections with meropenem or with piperacillin-tazobactam in the Intensive Care Burn Unit (ICBU) are shown in Table 2, Table 3, respectively, including detailed pharmacokinetic data and coverage against Gram-negative strains; creatinine clearance was estimated Cockcroft-Gault equation considering the ideal body weight. Patients were allocated in two groups based on the duration of infusion for both antimicrobial agents. At the last decade, pharmacokinetics of meropenem and piperacillin was based in serum levels after intermittent 0.5 hr.-infusion or extended 3-4 hrs.-infusion, or even the continuous infusion considered by many authors. Different results are related mainly to critically ill ICU patients included in clinical protocols mainly at the earlier or even at the late phase of septic shock. Pharmacokinetic-pharmacodynamics approach-based on serum levels to attain therapeutic target of $100\%f\Delta T > MIC$ for piperacillin and meropenem, was previously reported (Abdul-Aziz et al., 2015). Since the inflammatory response syndrome occurs by cytokine storm, mainly in the early phase of septic shock in ICU critically patients, the periods for blood sampling at the steady state for TDM, PK studies and drug effectiveness must be considered as reported for meropenem recently by comparison of pharmacokinetic data between the early *versus* late stage of therapy of septic shock in burn patients (Messiano et al., 2022). Serial of data related to PK/PD approach-based on serum levels was reported by many authors considering not only dose size and dose regimen, duration of infusion and renal function dependent, but also whether the protocol was conducted in ICU patients undergoing antibiotic therapy of nosocomial infections caused by Gram-negative strains, in the early stage, or even in late stage of septic shock, once significant PK changes occur impacting the coverage of meropenem and piperacillin. (Abdul-Aziz et al., 2016; Abdulla et al., 2020; Adnan et al., 2013; Camargo et al., 2022; Carlier et al., 2015; Chung et al., 2015; De Waele et al., 2014; Fournier et al., 2015; Gonçalves Pereira et al., 2015; Jeon et al., 2014; Messiano et al., 2022; Monti et al., 2023; Roberts et al., 2013; Santos et al., 2011; Santos et al., 2022; Santos et al., 2023a; Santos et al., 2023b;

Schoenenberger-Arnaiz et al., 2019; Sime et al., 2015; Silva Junior et al., 2017; Souza et al., 2019; Taccone et al., 2010).

Clinical Meropenem Study

Meropenem therapy was prescribed at dose regimen recommended in hospital for critically ill burn adults' patients with renal function preserved or augmented by vasopressors requirements. Data obtained in study protocol were related to characteristics of patients, all clinical and laboratorial data, coverage and outcome is detailed, Table 2.

Table 2. Burn septic patients undergoing meropenem therapy, median (IQR). Demographic, clinical, and laboratorial data, PK data, coverage, outcome

Meropenem 1g q8h (N = 25)	0.5 hr. -infusion n = 12	3 hrs. -infusion n = 13	P
Demographic data			
Gender (19M/6F)	9/3	10/3	1.000*
Age (yrs)	24(20-30)	28(24-39)	0.1817
Body weight (kg)	70(62-74)	70(60-75)	0.8273
Heights (cm)	171 (166-177)	169 (161-175)	0.7756
Body surface area (m^2)	1.82(1.58-1.89)	1.85(1.60-2.00)	0.8123
Body mass index (kg/m^2)	24(22-25)	23(22-24)	0.7647
Admission data			
SAPS3 higher than 57	62 (62-63)	61 (59-62)	0.2931
Risk of death (%)	24(16-36)	35(25-45)	0.3578
TBSA higher than 40 (%)	43(42-55)	51(49-52)	0.8454
Thermal/Electrical (19/6)	11/1	8/5	0.1602*
Inhalation injury (16/25)	8/12	8/13	0.1602*
Mechanical ventilation (16/25)	8/12	8/13	0.1602*
Vasopressors (21/25)	10	11	0.8167
Biomarkers-TDM (CSLI)			
C- reactive protein (mg/L)	224(179-286)	300(264-339)	0.7685
Leucocytes ($*10^3$ cells/mm^3)	9.34(4.66-17.68)	16.05(10.47-19.10)	0.7793
Neutrophils ($*10^3$ cells/mm^3)	8.55(2.40-15.22)	12.28(8.91-16.77)	0.8678
Creatinine clearance (mL/min)	120(95-138)	138(96-176)	0.8243
Dose normalized	1g q8h	1g q8h	
Daily dose (mg/kg)	44 (41-50)	43 (40-50)	0.8916
Dose regimen (mg/kg)	15 (14-17)	14(13-17)	0.9348
Pharmacokinetic data			
Biological half-life (hrs.)	2.0(1.8-2.1)	3.6(3.3-4.4)	0.0001
Total body clearance (L/h)	5.3(5.2-5.4)	8.6(8.2-8.9)	0.0001

Meropenem 1g q8h (N = 25)	0.5 hr. -infusion n = 12	3 hrs. -infusion n = 13	P
Pharmacokinetic data			
Volume of distribution (L)	15.4(14.1-16.2)	41.8(39.9-44.5)	0.0001
Microbiology (CSLI)	Coverage (patients)	Coverage (patients)	
MIC 1 mg/L	100% (12/12)	100% (13/13)	0.0001
MIC 2 mg/L	100% (12/12)	100% (13/13)	0.0001
MIC 4 mg/L	25% (3/12)	100% (13/13)	0.0001
MIC 8 mg/L	0% (0/12)	38% (5/13)	0.0001
Clinical outcome			
ICU period (days)	63 (55-75)	32 (28-35)	0.0001
Length of stay in hospital (days)	71 (63-81)	36 (33-41)	0.0001
Survivals	11/12	13/13	1.000*

Abbreviations: SAPS3: Simplified Acute Physiology Score III; TBSA: total burn surface area, Clcr: creatinine clearance; TDM: therapeutic drug monitoring; IQR: interquartile (25-75); ICU: Intensive care unit; LOS: length of stay; MIC: minimum inhibitory concentration; CSLI: Clinical Standard Laboratory Institute database.

Statistics: GraphPad Prism software, version 9.1.4, Mann Whitney, *Fisher test.

Twenty-five ICU septic major burn patients of both genders(19M/6F), at the earlier stage of septic shock, were included in the study to investigate antimicrobial effectiveness of therapy with meropenem 1 g q8h, dose regimen. Twelve septic burn patients received by intermittent 0.5 hr.-infusion: meropenem 1 g q8h (n = 12), and another thirteen patients received the same dose regimen by the new strategy recommended of an extended 3 hrs.-infusion, in group 2, n = 13 patients. Any significant difference was found between groups of patients included in meropenem clinical study related to demographic characteristics, clinical data at admission, laboratorial data, and dose regimen at TDM. On the other hand, difference between groups was pointed out to pharmacokinetic data, target attainment by PK/PD approach, and outcome. It is important to highlight that a very good linear correlation ($r^2 = 0.9871$, $P < 0.0001$) was stablished between increase on volume of distribution and a prolongation of biological half-life after meropenem dose regimen done by 3 hrs.-infusion. It is important to emphasize that the pharmacokinetics of meropenem reference data in healthy volunteers was dependent of the duration of infusion, as recently described in a review study done by Santos et al., (2023). This fact reported initially by Yin & Wang (2014), and more recently discussed by Lapczuk-Romanska et al., (2023) can be justified, based on the meropenem pharmacokinetics disposition that depends on drug elimination that occurs through glomerular filtration plus the tubular secretion carried out through the influx biotransporters OAT1-OAT3, and MRP4, the efflux biotransporters in the tubular cell of the renal proximal tubule. Then, total body clearance reduced for the carbapenem agent after both infusions by comparison with reference data reported in healthy volunteers is due to an inhibition of OAT1-OAT3 influx drug transporter into renal proximal tubular cell during the inflammatory cytokines storm that occurs mainly at the earlier stage of septic shock. Concerning the meropenem coverage, pharmacodynamics was investigated by isolating pathogens from fluids and secretions collected from septic burn patients considering the total number of isolates from the site of infection. It should be mentioned that in this evaluation the same patient could contribute to the count of infection in more than one site of infection according to microbiological colonization. A total of 37 isolates were stratified into Gram-negative susceptible pathogens-Enterobacteriaceae (18 isolates) including *K. pneumoniae* up to MIC 2 mg/L, and non-Enterobacteriaceae (19 isolates) including *B. cepaceae, P. aeruginosa and A. baumannii*. On the other hand, for patients undergoing meropenem therapy by both infusions it was isolated two *Burkolderia cepaceae,* ten *Pseudomonas aeruginosa,* meropenem susceptible up to MIC 2 mg/L, and eight *Acinetobacter baumannii,* resistant to meropenem (MIC > 16 mg/L). So, for these patients colistin was included in the therapy until negative cultures. Superiority of meropenem by extended 3 hrs.-infusion related to efficacy-safety on antimicrobial coverage over the intermittent 0.5 hr.-infusion was demonstrated. Based on this strategy, clinical and microbiological cure occurred of all patients in a short period after the extended 3 hrs.- infusion.

Clinical Piperacillin Study

Demographic, clinical characteristics of patients at admission, and during drug serum monitoring of them undergoing therapy of infections with piperacillin-tazobactam in the Intensive Care Burn Unit (ICBU) are shown in Table 3; creatinine clearance was estimated Cockcroft-Gault equation. Piperacillin therapy was prescribed at dose regimen recommended in our hospital for critically ill burn adults' patients with renal function preserved or augmented by vasopressors requirements. Twenty ICU septic major burn patients of both genders (11M/9F), at the earlier stage of septic shock, were included in the study to investigate effectiveness of therapy with piperacillin-tazobactam 4.5 g q6h dose regimen. Patients were allocated in two groups based on the duration of infusion for both antimicrobial agents. Ten septic burn patients received by intermittent 0.5 hr.-infusion: piperacillin-tazobactam 4.5 g q6h (n = 10), and another ten patients received the same dose regimens of both antimicrobial agents by the new strategy recommended of an extended 3 hrs.-infusion, in group 2, n = 10 patients. Blood was sampling 48 hrs. after the therapy started for drug serum measurements by LC-UV at the 3rd hr. of started infusion, and one hour before the next infusion. Patients' drug serum samples were analyzed by liquid chromatography with detection set at 210 nm. Pharmacokinetics of piperacillin in septic major burns was based on noncompartmental data analysis to estimate biological half-life, total body clearance, and volume of distribution at the steady state, and data reported in healthy volunteers for piperacillin pharmacokinetics were considered. Thus, kinetic parameters obtained in burn patients were compared with data reported in healthy volunteers. Thus, changes on pharmacokinetics in severely burned patients after intermittent or extended infusion were compared with the reference data described in healthy volunteers by Occhipinti et al., (1993). It is important to emphasize that the pharmacokinetics reference data in healthy volunteers for piperacillin were independent of the duration of infusion. Consequently, the reference values for PK parameters of piperacillin in healthy volunteers are independent of the duration of infusion. Non-parametric statistics showed a difference between groups related to a two-fold prolongation of the biological half-life and an increase by twice in the apparent volume of distribution for septic patients who received a 3-hrs.- infusion compared to patients who received an intermittent infusion of 0.5 hr. A linear correlation ($r^2 = 0.8742$, $p < 0.0001$) was registered between both parameters after the 3 hrs.-infusion. Pharmacokinetic-pharmacodynamics approach-based on serum levels to attain therapeutic target of 100%$f\Delta T >$ MIC was considered. Drug effectiveness was evaluated by pharmacokinetic-pharmacodynamics approach, guided by PK/PD target attainment and microbiology of Gram-negative isolates from cultures. Piperacillin coverage occurred for all patients receiving extended infusion; while it was found that piperacillin coverage against Gram-negative isolates occurred only up to MIC 2 mg/L. It should be mentioned that in this evaluation the same patient could contribute to the count of infection in more than one site of infection according to microbiological colonization. A total of 25 isolates were stratified into Gram-negative susceptible pathogens-Enterobacteriaceae (20 isolates) up to MIC 2 mg/L including *Escherichia coli, Enterobacter cloacae, Haemophylus influenza, Morganela morganii, Proteus mirabilis, Providencia stuartii, Serratia marcences, K. pneumoniae* in patients with septic burns receiving piperacillin 4.5 g q6h by intermittent or extended infusion; non-Enterobacteriaceae (5 isolates) included only *P aeruginosa*. Consequently, the superiority of piperacillin extended 3 hrs.-infusion on antimicrobial coverage over the intermittent 0.5 hr.-infusion was demonstrated.

Table 3. Burn septic patients undergoing piperacillin therapy, median (IQR). Demographic, clinical at admission, laboratorial data, PK data and coverage

Piperacillin-Tazobactam 4.5g q6h (N = 20)	0.5 hr. -infusion n = 10	3 hrs. -infusion n = 10	P
Demographic data			
Gender (11M/9F)	6M/4F	5M/5F	0.4018*
Age (yrs)	30 (24-31)	27 (24-34)	0.9620
Body weight (kg)	70 (57-75)	73 (55-79)	0.5986
Heights (cm)	171(166-177)	173 (165-179)	0.5983
Body surface area (m^2)	1.82 (1.58-1.89)	1.87 (1.60-2.00)	0.9397
Body mass index (kg/m^2)	24 (22-25)	24 (20-25)	0.9369
Admission data			
SAPS3	54 (49-60)	48 (38-60)	0.1677
Risk of death (%)	24(16-36)	15(5-35)	0.2861
TBSA (%)	30(20-42)	34(18-38)	0.9874
Thermal/Electrical (17/3)	9/1	8/2	1.0000*
Inhalation injury (16/20)	8/10	8/10	1.0000*
Mechanical ventilation (16/20)	8/10	8/10	1.0000*
Vasopressors (18/20)	8	10	0.5434*
Biomarkers-TDM (CSLI)			
C- reactive protein (mg/L)	232 (107 – 285)	183 (118 – 279)	0.9747
Leucocytes (*10^3 cells/mm^3)	10.91(4.60-13.93)	14.05 (11.47-18.10)	0.3025
Neutrophils (*10^3 cells/mm^3)	8.55 (3.40 – 11.22)	10.28 (8.91-15.77)	0.5764
CLcr (mL/min)	115 (83-148)	127 (90-170)	0.6463
Dose normalized	4.5g q6h	4.5g q6h	
Daily dose (mg/kg)	229 (217-297)	225 (214-246)	0.5620
Dose regimen (mg/kg)	57 (54-74)	62 (51-72)	0.6236
Pharmacokinetic data			
Elimination half-life (hrs.)	1.1 (1.0-1.2)	1.9(1.7-2.5)	0.0004
Total body clearance (L/h)	8.7 (8.0-9.0)	11.0 (10.0-14.0)	0.0085
Volume of distribution (L)	14.0 (13.0-15.0)	34.0 (23.0-44.0)	0.0001
Microbiology (CSLI)	Coverage (patients)	Coverage (patients)	
MIC 1 mg/L	100% (10/10)	100% (10/10)	0.0001
MIC 2 mg/L	100% (10/10)	100% (10/10)	0.0001
MIC 8 mg/L	0% (0/10)	100% (10/10)	0.0001
MIC 16 mg/L	0% (0/10)	100% (10/10)	0.0001

Abbreviations: SAPS3: Simplified Acute Physiology Score III; TBSA: total burn surface area, CLcr: creatinine clearance; TDM: therapeutic drug monitoring; IQR: interquartile (25-75); MIC: minimum inhibitory concentration; CSLI: Clinical Standard Laboratory Institute database.

Statistics: GraphPad Prism software, version 9.1.4, Mann Whitney for unpaired data, *Fisher test.

Based on this strategy, clinical and microbiological cure occurred of all patients in a short period after the extended 3 hrs.- infusion. All data obtained in Piperacillin-Tazobactam study protocol were related to characteristics of patients including clinical and laboratorial data, pharmacokinetics, and target attainment by PK/PD approach, as detailed in Table 3.

Conclusion

Superiority of liquid chromatographic bioanalytical methods developed and validated were shown based on large linearity and lower limits of quantification and detection by 5 μL injected volume, with acceptable intraday/inter-day precisions and systematic error. Good stability of standards on the bench (4 hrs.) for both analytes in serum samples, short stability on the bench (6 hrs.), post-process stability of serum extracts up to 24 hrs. in the rack of autosampler, serum samples long stability in ultralow freezer storage, as well as cycles freezing/de-freezing stability, and the robustness of chromatographic system.

Concerning therapy of septic shock with meropenem in burn adults against Gram-negatives strains, the superiority of 3 hrs.-infusion over 0.5 hr.-infusion was registered. Clinical cure occurred for G1-patients just against Gram-negative susceptible strains up to MIC 2 mg/L after the intermittent infusion, while clinical cure occurred for all G2-patients including against strains of intermediate susceptibility up to MIC 4 mg/L and it was increased up to MIC 8 mg/L in 5/13 (38%) patients after the extended infusion. It is important to highlight that difference between groups was pointed out as a function of duration of infusion that changes the meropenem pharmacokinetics related to biological half-life prolonged, because of increases on volume of distribution that impacted positively the coverage, even with vasopressor requirements during the systemic inflammatory response syndrome.

In addition, piperacillin-tazobactam therapy also was guided by the cultures, and the target was attained by PK/PD approach based on serum levels by TDM done in a real time. Concerning therapy of septic shock in burn adults against Gram-negatives strains, the superiority of 3 hrs.-infusion over 0.5 hr.-infusion was also registered. Clinical cure occurred for all G1-patients against strains just up to MIC 2 mg/L, while the coverage for G2-patients was guaranteed up to MIC 16 mg/L. It is important to highlight that piperacillin pharmacokinetics depends mainly on glomerular filtration rate and vasopressor requirements. Difference between groups was pointed out as a function of duration of infusion by the significant increases on piperacillin systemic availability after the extended 3 hrs.- infusion, that impacted positively the antimicrobial coverage in those patients, even with vasopressor requirements during the systemic inflammatory response syndrome.

Finally, PK/PD approach done in real-time based on drug serum monitoring by LC-UV is an important tool to assess soon drug effectiveness-safety in septic major burns, especially during the systemic inflammatory response syndrome that occurs in these patients at the early stage of septic shock. Consequently, antimicrobial therapy against Gram-negative nosocomial pathogens for the combat of microbial resistance must be considered by applying bioanalytical methods based on liquid chromatography, that is quite selective for antimicrobial TDM serum levels done in a real time.

Disclaimer

None.

References

Abdul-Aziz M. H., Lipman J., Akova M., Bassetti M., De Waele J. J., Dimopoulos G., Dulhunty J., Kaukonen K. M., Koulenti D., Martin C., Montravers P., Rello J., Rhodes A., Starr T., Wallis S. C., Roberts J. A., DALI Study Group. Is prolonged infusion of piperacillin/tazobactam and meropenem in critically ill patients associated with improved pharmacokinetic/pharmacodynamic and patient outcomes ? An observation from the Defining Antibiotic Levels in Intensive care unit patients. *DA. Journal of Antimicrobial Chemotherapy*, v. 71, n. 1, p. 196–207, 2016. Available in: http://doi.org/10.1093/jac/dkv288.

Abdul-Aziz M. H., Lipman, J., Mouton, J. W., Hope, W. W., Roberts, J. A. Applying pharmacokinetic/pharmacodynamic principles in critically ill patients: optimizing efficacy and reducing resistance development. *Semin Respir Crit Care Med.* 36(1):136–53, 2015. Available in: http://dx.doi.org/10.1055/s-0034-1398490.

Abdulla A., Ewoldt T. M. J., Hunfeld N. G. M., Muller A. E., Rietdijk W. J. R., Polinder S., van Gelder T., Endeman H., Koch B. C. P. The effect of therapeutic drug monitoring of beta-lactam and fluoroquinolones on clinical outcome in critically ill patients: the DOLPHIN trial protocol of a multi-Centre randomized controlled trial. *BMC Infectious Diseases* (2020). Available in: https://doi.org/10.1186/s12879-020-4781-x.

Adnan S., Li J. X., Wallis S. C., Rudd M., Jarrett P., Paterson D. L., Lipman J., Udy A. A., Roberts J. A. Pharmacokinetics of meropenem and piperacillin in critically ill patients with indwelling surgical drains. *International Journal of Antimicrobial Agents*. v.

42(1): p. 90–93, 2013. Available in: http://dx.doi.org/10.1016/j.ijantimicag.2013.02.023.

Barlam T. F., Cosgrove S. E., Abbo L. M., MacDougall C., Schuetz A. N., Septimus E. J., Srinivasan A., Dellit T. H., Falck-Ytter Y. T., Fishman N. O., Hamilton C. W., Jenkins T. C., Lipsett P. A., Malani P. N., May L. S., Moran G. J., Neuhauser M. M., Newland J. G., Ohl C. A., Samore M. H., Seo S. K., Trivedi K. K. Implementing an Antibiotic Stewardship Program: Guidelines by the Infectious Diseases Society of America and the Society for Healthcare Epidemiology of America. *Clinical Infectious Diseases*. v. 62. n. 10. p. e51–e77. 13 April 2016. Available in: http://dx.doi.org/10.1093/cid/ciw118.

Cain S. E., Kohn J., Bookstaver P. B., Albrecht H., Al-Hasan M. N. Stratification of the Impact of Inappropriate Empirical Antimicrobial Therapy for Gram-Negative Bloodstream Infections by Predicted Prognosis. *Antimicrobial Agents and Chemotherapy*. 59(1):245–250 (2015). Available in: https://doi.org/10.1016/j.burns.2010.06.005.

Carlier M., Stove V., Wallis S. C., De Waele J. J., Verstraete A. G., Lipman J., Roberts J. A. Assays for Therapeutic Drug Monitoring of β-Lactam Antibiotics: A Structured Review. *International Journal of Antimicrobial Agents*. v. 46. n. 4. p. 367–375. October 2015. Available in: http://dx.doi.org/10.1016/j.ijantimicag.2015.06.016.

Caro Y. S., Cámara M. S., De Zan M. M. A review of bioanalytical methods for therapeutic drug monitoring of beta-lactam antibiotics in critically ill patients: Evaluation of the approaches used to develop and validate quality attributes. *Talanta* 210 (2020) 120619. Available in: https://doi.org/10.1016/j.talanta.2019.120619.

Casals G., Hernández C., Hidalgo S., Morales B., López-Púa Y., Castro P., Fortuna V., Martínez J. A., Brunet M. Development and validation of a UHPLC diode array detector method for meropenem quantification in human plasma. *Clin. Biochem*. v.47, p.223–227 (2014). Available in: https://doi.org/10.1016/j.clinbiochem.2014.08.002.

Casals G., Hernández C., Hidalgo S., Morales B., López-Púa Y., Castro P., Fortuna V., Martínez J. A., Brunet M. Development and validation of a UHPLC diode array detector method for meropenem quantification in human plasma. *Clin. Biochem*. 47: 223–227 (2014). Available in: http://dx.doi.org/10.1002/jcph.505.

De Camargo T. V., Junior, E. M. S., Silva Jr, J. M., Gomez, D. S. & Santos, S. R. C. J. PK/PD approach to evaluate Meropenem effectiveness in critically ill burn adolescents versus young adults undergoing therapy of septic shock. *Pharm Pharmacol Int J*. 10(3), 79-85 (2022). https://doi.org/10.15406/ppij.2022.10.00368.

De Waele J. J., Carrette S., Carlier M., Stove V., Boelens J., Claeys G., Leroux-Roels I., Hoste E., Depuydt P., Decruyenaere J., Verstraete A. G. Therapeutic Drug Monitoring-Based Dose Optimisation of Piperacillin and Meropenem: A Randomised Controlled Trial. *Intensive Care Medicine*. v. 40(3) p. 380–387. December 2014. Available in: http://dx.doi.org/10.1007/s00134-013-3187-2.

Evans L., Rhodes A., Alhazzani W., Antonelli M., Coopersmith C. M., French C., Machado F. R., Mcintyre L., Ostermann M., Prescott H. C., Schorr C., Simpson S., Wiersinga W. J., Alshamsi F., Angus D. C., Arabi Y., Azevedo L., Beale R., Beilman G., Belley-Cote E., Burry L., Cecconi M., Centofanti J., Coz Yataco A., De Waele J., Dellinger R. P., Doi K., Du B., Estenssoro E., Ferrer R., Gomersall C., Hodgson C., Møller M.

H., Iwashyna T., Jacob S., Kleinpell R., Klompas M., Koh Y., Kumar A., Kwizera A., Lobo S., Masur H., McGloughlin S., Mehta S., Mehta Y., Mer M., Nunnally M., Oczkowski S., Osborn T., Papathanassoglou E., Perner A., Puskarich M., Roberts J., Schweickert W., Seckel M., Sevransky J., Sprung C. L., Welte T., Zimmerman J., Levy M. Surviving Sepsis Campaign: International Guidelines for Management of Sepsis and Septic Shock 2021. *Critical Care Medicine*: November 2021 - Volume 49 – Issue 11 - p 1063-1143. Available in: http://dx.doi.org/10.1097/CCM.0000000000005337.

Ferrone V., Cotellese R., Di Marco L., Bacchi S., Carlucci M., Cichella A., Raimondi P., Carlucci G. Meropenem, levofloxacin and linezolid in human plasma of critical care patients: a fast semi-automated micro-extraction by packed sorbent UHPLC-PDA method for their simultaneous determination, *J. Pharm. Biomed. Anal.* 140 : 266–273 (2017). Available in: https://doi.org/10.1016/j.jpba.2017.03.035.

Fournier A., Eggimann P., Pagani J. L., Revelly J. P., Decosterd L. A., Marchetti O., Pannatier A., Voirol P., Que Y. A. Impact of the introduction of real-time therapeutic drug monitoring on empirical doses of carbapenems in critically ill burn patient. *Burns* 41 (2015). Available in: https://dx.doi.org/10.1016;j.burns.2025.01.001.

Global Sepsis Alliance. 4th World Sepsis Congress on April 25 and 26, 2023. One Global Health Treat: Sepsis, Pandemics, and Antimicrobial Resistance. Available in: https://www.worldsepsiscongress.org.

Global Sepsis Alliance. Sepsis and COVID-19/Coronavirus/SARS-COV-2, 2020. Available in: https://www.global-sepsis-alliance.org/covid19.

Goncalves-Pereira J., Silva N. E., Mateus A., Pinho C., Povoa P. Assessment of pharmacokinetic changes of meropenem during therapy in septic critically ill patients. *BMC Pharmacol Toxicol.* v. 15:21, April 14. (2014). Available in: https://doi:10.1186/2050-6511-15-21. http://www.biomedcentral.com/2050-6511/15/21.

Greenhalgh D. G., Saffle J. R., Holmes J. H. 4th, Gamelli R. L., Palmieri T. L., Horton J. W., Tompkins R. G., Traber D. L., Mozingo D. W., Deitch E. A., Goodwin C. W., Herndon D. N., Gallagher J. J., Sanford A. P., Jeng J. C., Ahrenholz D. H., Neely A. N., O'Mara M. S., Wolf S. E., Purdue G. F., Garner W. L., Yowler C. J., Latenser B. A. American burn association consensus conference to define sepsis and infection in burns. *Journal of Burn Care & Research.* 28(6):776–790 (2007). Available in: https://doi.org/10.1097/BCR.0b013e3181599bc9.

Jaruratanasirikul S., Sriwiriyajan S. Comparison of the Pharmacodynamics of Meropenem in Healthy Volunteers Following Administration by Intermittent Infusion or Bolus Injection. *Journal of Antimicrobial Chemotherapy.* v. 52(3): p. 518–521 (2023). Available in: https://doi.org/.1093/jac/dkg378.

Jeon S., Han S., Lee J., Hong T., Paek J., Woo H., Yim D. S. Population Pharmacokinetic Analysis of Piperacillin in Burn Patients. *Antimicrobial Agents and Chemotherapy.* v. 58(7). p. 3744–3751. 21 April 2014. Available in: http://dx.doi.org/10.1128/AAC.02089-13.

Łapczuk-Romańska J., Droździk M., Oswald S., Droździk M. Drug Transporters in Pharmacotherapy *Int. J. Mol. Sci.* v. 24: p. 28-56 (2023). Available in: https://doi.org/10.3390/ijms24032856.

McWhinney B. C., Wallis S. C., Hillister T., Roberts J. A., Lipman J., Ungerer J. P. Analysis of 12 beta-lactam antibiotics in human plasma by HPLC with ultraviolet detection. *J. Chromatogr.* B, v. 878: p. 2039–2043 (2010). Available in https://doi.org/10.1016/j.jchromb.2010.05.027.

Messiano C. G., Junior, R. M., Pereira, G. O., da Silva Junior, E. M., Gomez, D. S. & Santos, S. R. C. J. Therapeutic Target Attainment of 3-Hour Extended Infusion of Meropenem in Patients with Septic Burns. *Clinical Therapeutics*, 44(4), 624-629 (2022). Available in: http://dx.doi.org/10.1016/j.clinthera.2022.02.007.

Monti G., Bradic N., Marzaroli M., Konkayev A., Fominskiy E., Kotani Y., Likhvantsev V. V., Momesso E., Nogtev P., Lobreglio R., Redkin I., Toffoletto F., Bruni A., Baiardo Redaelli M., D'Andrea N., Paternoster G., Scandroglio A. M., Gallicchio F., Ballestra M., Calabrò M. G., Cotoia A., Perone R., Cuffaro R., Montrucchio G., Pota V., Ananiadou S., Lembo R., Musu M., Rauch S., Galbiati C., Pinelli F., Pasin L., Guarracino F., Santarpino G., Agrò F. E., Bove T., Corradi F., Forfori F., Longhini F., Cecconi M., Landoni G., Bellomo R., Zangrillo A. Continuous vs Intermittent Meropenem Administration in Critically Ill Patients with Sepsis. The MERCY Randomized Clinical Trial. *JAMA*. https://doi.org/10.1001/jama.2023.10598.

Nilsson-Ehle I., Hutchison M., Haworth S. J., Norrby S. R. Pharmacokinetics of meropenem compared to imipenem-cilastatin in young healthy males. *Eur. J. Clin. Microbiol. Infect. Dis.* v. 10(2): p. 85–88 (1991). Available in: https://doi.org/10.1007/BF01964413.

Occhipinti D. J., Pendland S. L., Schoonover L. L., Rypins E. B., Danziger L. H., Rodvold K. A. Pharmacokinetics and pharmacodynamics of two multiple-dose piperacillin-tazobactam regimens. *Antimicrobial Agents Chemotherapy*. v. 41(11): p. 2511-7, November 1997. Available in: https://pubmed.ncbi.nlm.nih.gov/9371358/.

Roberts J. A., Hope W. W., Lipman J. Therapeutic Drug Monitoring of β-Lactams for Critically Ill Patients: Unwarranted or Essential? *International Journal of Antimicrobial Agents*. v. 35(5): p. 419–420. May 2010. Available in: http://dx.doi.org/10.1016/j.ijantimicag.2010.01.022.

Roth T., Fiedler S., Mihai S., Parsch H. Determination of meropenem levels in human serum by high-performance liquid chromatography with ultraviolet detection. *Biomed. Chromatogr.* v. 31 (2017). Available in: https://doi.org/10.1002/bmc.3880.

Rudd K. E., Johnson S. C., Agesa K. M., Shackelford K. A., Tsoi D., Kievlan D. R., Colombara D. V., Ikuta K. S., Kissoon N., Finfer S., Fleischmann-Struzek C., Machado F. R., Reinhart K. K., Rowan K., Seymour C. W., Watson R. S., West T. E., Marinho F., Hay S. I., Lozano R., Lopez A. D., Angus D. C., Murray C. J. L., Naghavi M. Global. Regional. and National Sepsis Incidence and Mortality. 1990–2017: Analysis for the Global Burden of Disease Study. *The Lancet*. v. 395(10219): p. 200–211. January 2020. Available in: http://dx.doi.org/10.1016/S0140-6736(19) 32989-7.

Santos S. R. C. J., de Camargo T. V., Macedo V. A., Souza, V. K. & Gomez, D. S. Vancomycin and meropenem serum monitoring for target attainment by PK/PD approach as an effective tool in the battle against nosocomial pathogens in septic pediatric burn patients. *Pharm Pharmacol Int J.* 10(3):87–95 (2022). Available in: https://doi.org/10.15406/ppij.2022.10.00369.

Santos S. R. C. J., de Camargo T. V., Messiano C. G., Kupa L. V. K., Souza V. K., Ronaldo Morales Jr. R., Sanches Pinto D. C. S., Silva Junior E. M., Silva Junior J. M., Gomez D. S. Combating Bacterial Resistance to Antimicrobials in Severe Septic ICU Patients: Importance of Meropenem. Piperacillin Serum Monitoring as a Dose Adjustment and Duration of Infusion Strategies. *Pharm. Pharmacol. Int. J.* v. 11(2): p. 52–60 (2023a), May 24. Available in: https://doi.org/10.15406/ppij.2023.11.00402.

Santos, S. R. C. J., de Camargo T. V., Messiano C. G., Souza V. K. Frederico Ribeiro Pires, Débora Cristina Sanches Pinto, Edvaldo Vieira de Campos, Elson Mendes da Silva Junior, João Manoel da Silva Junior, David de Souza Gomez. Combined vancomycin-meropenem therapy in pediatric major burns undergoing therapy of septic shock guided by cultures and pharmacokinetic-pharmacodynamics approach based on serum levels to combat bacterial resistance. *Pharm Pharmacol Int J.* 11(2):62–68 (2023b). May 24. Available in: https://doi.org/10.15406/ppij.2023.11.00403.

Santos S. R. C. J., Sanches-Giraud, C., De Souza, F. F., Gómez, D. S., Campos, E. V., Azevedo, R., Ferreira, M., Nascimento, J. Pharmacokinetic-pharmacodynamic correlation for meropenem applied to a burn child using a bioanalytical liquid chromatographic method. *Revista Portuguesa de Farmacoterapia.* v. 3(4): p. 224–232. October 2011. Available in: https://doi.org/10.25756/rpf.v3i4.79.

Schoenenberger-Arnaiz J. A., Ahmad-Diaz F., Miralbes-Torner M., Aragones-Eroles A., Cano-Marron M., Palomar-Martinez M. Usefulness of Therapeutic Drug Monitoring of Piperacillin and Meropenem in Routine Clinical Practice: A Prospective Cohort Study in Critically Ill Patients. *European Journal of Hospital Pharmacy.* v. 27(e1): p.e30–e35. 27, February 2019. Available in: http://dx.doi.org/10.1136/ejhpharm-2018-001713.

Silva Jr J. M., Oliveira A. M. R. R., Campos E. V., Souza V. K., Kupa, L. V. K., Gomides A. S., Ferreira G. A., Oliveira T. C., Gomez D. S., Santos, S. R. C. J. Piperacillin effectiveness in septic burn patients by comparison of two empiric daily dose 12 *versus* 16 g against susceptible strains based on drug plasma measurements done in a real time. *Critical Care.* v. 21(2): p. 31-31, July 2017. Available in: http://dx.doi.org/10.1186/s13054-017-1706-1.

Sime F. B., Roberts M. S., Tiong I. S., Gardner J. H., Lehman S., Peake S. L., Hahn U., Warner M. S., Roberts J. A. Can Therapeutic Drug Monitoring Optimize Exposure to Piperacillin in Febrile Neutropenic Patients with Haematological Malignancies? A Randomized Controlled Trial. *Journal of Antimicrobial Chemotherapy.* v. 70 (8): p. 2369–2375. May 2015. Available in: http://dx.doi.org/10.1093/jac/dkv123.

Souza V. K., Silva Jr E. M., Kaneko V. M., Silva Jr J. M., Gomides A. S., Ferreira G. A., Oliveira T. C., Gomez D. S., Santos, S. R. C. J. Extended two hours infusion improves piperacillin effectiveness against susceptible strains in septic burn patients. *Critical Care.* v.23(3): p 21-22 (2019). https://doi.org/10.1186/s13054-019-9. ISSN 1364-8535.

Taccone F. S., Laterre P. F., Dugernier T., Spapen H., Delattre I., Wittebole X., De Backer D., Layeux B., Wallemacq P., Vincent J. L., Jacobs F. Insufficient β-Lactam Concentrations in the Early Phase of Severe Sepsis and Septic Shock. *Critical Care.* v.14(4): p. R126 (2010). Available in: http://dx.doi.org/10.1186/cc9091.

Udy A. A., Lipman J., Jarrett P., Klein K., Wallis S. C., Patel K., Kirkpatrick C. M., Kruger P. S., Paterson D. L., Roberts M. S., Roberts J. A. Are Standard Doses of Piperacillin Sufficient for Critically Ill Patients with Augmented Creatinine Clearance? *Critical Care*. v. 19(1): p. 28. 2015. Available in: http://dx.doi.org/10.1186/s13054-015-0750-y.

World Health Organization. From emergency response to long-term covid-19 disease management: Sustaining gains made during the COVID-19 pandemic. May 4th, 2023. Available in: https://www.who.int/publications/i/item/WHO-WHE-SPP-2023.1.

Yin J., Wang J. Yin J., Wang J. Renal drug transporters and their significance in drug-drug interactions. *Acta Pharm Sin*. v. 6(5): p. 363-373, September 6, 2016. Available in: http://dx.doi.org/10.1016/j.apsb.2016.07.013.

Zou L., Meng F., Hu L., Huang Q., Liu M., Yin T. A novel reversed-phase high-performance liquid chromatographic assay for the simultaneous determination of imipenem and meropenem in human plasma and its application in TDM. *J. Pharm. Biomed. Anal.* 169:142–150 (2019). Available in: https://doi.org/10.1016/j.jpba.2019.01.039.

Biographical Sketches

Professor Silvia Regina Cavani Jorge Santos
(Professor of Clinical Pharmacokinetics)
Leadership 1 of Research Team

ORCID https://orcid.org/0000-0001-5295-2079 Brazilian Platform of Research (CNPq).

Affiliation:
University of São Paulo Division of Plastic Surgery & Burns of Clinics Hospital of Medical School (2007-2023). University of Sao Paulo, Clinical Pharmacokinetics Center, School of Pharmaceutical Sciences (1988-2023).

Education:
University of São Paulo Pharmacist, Toxicology expert (1973); Master of Toxicology (1976); Doctor of Pharmacy - Toxicology (PhD 1982) Pharmacokinetics/Expertise (sandwich grant-England 1980-81). Post doctoral studies in research Laboratories (Medical School of Essen (1986), Germany, Medical School of Kiel (1987), Germany; School of Clinical Pharmacology, University of Leiden & Leiden Amsterdam Center for Drug Research (1992), Netherlands; Medical School of Dresden (1994), Germany.

Research Experience:

41 Doctor and Master of Science Students, University of Sao Paulo: Graduate Students Advisor since 1988).

Membership of international Societies European Burn Association (EBA); American Society of Microbiology (ASM); American College of Clinical Pharmacy (ACCP).

Leadership of Research "Clinical Pharmacokinetics" Group of a team of 37 physicians, surgeons, intensivists, nurses, clinical pharmacists, also including the antimicrobial serum monitoring done by the laboratory team of hospital.

Professional Experience:

Expertise of our group at the last 20 years in a tertiary public hospital is focused on critically ill patients (bed side monitoring the septic shock) undergoing combined therapy of bacterial infections caused by Gram-positive and Gram-negative strains with antibiotics, or even combined with antifungal agents. Main goals of our team are to guarantee antimicrobial effectiveness and safety by individualization of therapy.

Professional Appointments:

1. Therapeutic Drug Monitoring Laboratory & Clinical Pharmacokinetics Director, Heart Institute, Clinical Hospital, Medical School University of Sao Paulo (1977-2004).
2. School of Pharmaceutical Sciences University of Sao Paulo Lecturer (1988).
3. School of Pharmaceutical Sciences University of Sao Paulo Associate Professor (1988-2000).
4. School of Pharmaceutical Sciences Professor full on Clinical Pharmacokinetics & Pharmacotherapy University of Sao Paulo (2001-2023)
5. School of Pharmaceutical Sciences University of Sao Paulo: Head of Pharmacy Department (2004-2010), Chairman of Committees: Ethics, Research, Graduate Studies, International Cooperation (1988-2022).

Honors:

Professor University of Sao Paulo; Fellowship of International Societies: European Burns Association (EBA), American Society of Microbiology (ASM), American College of Clinical Pharmacy (ACCP).

Scientific Prizes last three years: National and international scientific events:

Brazilian Society of Infectiology (BSI) December/2020. The Brazilian Journal of Inf. Diseases, Volume 25, Supplement 1, January 2021 ISSN 1413-86-70. https://doi.org/10.1016/j.bjid.2020.101387.

"Comparative study of the effectiveness of Piperacillin-Tazobactam after intermittent versus extended infusion in severe burn septic patients by the pharmacokinetic-pharmacodynamic approach (PK/PD)." Vanessa Kasubeck de Souza Master of Pharmacy Student, Advisor Silvia R. C. J. Santos.

Asian Pacific Foundation for Infectious Diseases "2021 Outstanding Poster Award" 13th International Symposium on Antimicrobial Agents and Resistance (ISAAR-2021). e-Poster Presentation Winner promoted by Asian Pacific Foundation for Infectious Diseases (APFID) September 9-10/2021. Apfid-Prize (Asian Pacific Foundation for Infectious Diseases) "2021 Leonard de Vinci Kanda Kupa. Ph.D. Toxicology-Clinical Pharmacpkinetics (2019) Pos-Doc (2019-2022). Published: *International Journal of Antimicrobial Agents,* Volume 58, Supplement 1, September 2021, 2100279-21002818 Morales, Jr, R., Kupa, L. V. K., Vianna, K. B., Garcia, M. S., Santos, V. J., Campos, E. V., Silva Jr, J. M., Silva Jr E. M., Oliveira, T. C., Gomez, D. S. & Santos, S. R. C. J. (2021). Meropenem extended infusion versus intermittent infusion against nosocomial MIC 4 mg/L strains to guarantee drug effectiveness by PK/PD approach in burn patients at the earlier period of septic shock. In *International Journal of Antimicrobial Agents.* Elsevier. https://doi.org/10.1016/j.ijantimicag.2021.106421.10.

III International Meeting Sobratafe on advanced wound care, 2019 Maksoud Plaza Hotel São Paulo, SP. Awards at the Event Promoter: Brazilian Society for Advanced Treatment of Complex Wounds (Sobratafe) Awards Sobratfe 2019.

Award 1: Meropenem effectiveness in critically ill burn patients during the septic shock by pharmacokinetics assessment based on drug serum measurements done in a real time. Claudia Garcia Messiano, Elson Mendes da Silva Júnior, João Manoel da Silva Júnior, Edvaldo Vieira de Campos, Amanda Maria Ribas Rosa de Oliveira, Aline S. Gomides, Gabriela A. Ferreira, Thiago C. Oliveira, Silvia Regina Cavani Jorge Santos, David de Souza Gomez. Claudia Garcia Messiano Master of Pharmacy Student, Advisor Silvia R. C. J. Santos.

Award 2: Vancomycin dose adjustment done in areal time for target attainment in critically ill septic pediatric burns versus non-burns against

Staphylococcus spp MIC 1 mg/L. Camila Salinas Guillaux, Ronaldo Morales Júnior, Vedilaine Aparecida Bueno da Silva Macedo, Elson Mendes da Silva Júnior, João Manoel da Silva Júnior, Edvaldo Vieira de Campos, Amanda Maria Ribas Rosa de Oliveira, Aline Sandré Gomides, Gabriela Aparecida Ferreira, Thiago C. Oliveira, Silvia Regina Cavani Jorge Santos, David de Souza Gomez. PhD Student Ronaldo Morales Jr, Advisor Silvia R. C. J. Santos.

National Meeting Sobratafe on Advanced wound Care, Sobratafe 2020. Maksoud Plaza Hotel São Paulo, SP. Promoted by the Brazilian Society for Advanced Treatment of Complex Wounds (Sobratafe) Dr Walter Soares Pinto Prize.

Award 1: Is Meropenem effectiveness in ICU burn patients during SIRS at the earlier period of septic shock impacted by pharmacokinetic changes? Adolescents versus Young adults. Thais V. de Camargo, Edvaldo Vieira de Campos, Amanda Maria Ribas Rosa de Oliveira, João Manoel da Silva Júnior, Elson Mendes da Silva Júnior, Aline S. Gomides, Gabriela A. Ferreira, Thiago C. Oliveira, Vera Lúcia Lanchote, Silvia Regina Cavani Jorge Santos, David de Souza Gomez. Thais Vieira de Camargo Master of Pharmacy Student, Advisor Silvia R. C. J. Santos.

Award 2: Extended infusion improves beta-lactams effectiveness in septic burn patients with normal renal function against P aeruginosa intermediate susceptibility. Vanessa Kasubeck de Souza, Karina B Vianna, João Manoel da Silva Júnior, Amanda Maria Ribas Rosa de Oliveira, Edvaldo Vieira de Campos, Elson Mendes da Silva Júnior, Aline S. Gomides, Gabriela A. Ferreira, Thiago C. Oliveira, Vera Lúcia Lanchote, Silvia Regina Cavani Jorge Santos, David de Souza Gomez. Vanessa Kasubeck de Souza Master of Pharmacy Student, Advisor Silvia R. C. J. Santos.

Award 3: Pharmacokinetic-pharmacodynamic approach permits vancomycin dose adjustment for effectiveness against infection caused by Gram-positive pathogens MIC > 1 mg/L in ICU paediatric burn patients. Ronaldo Morales Júnior, Thais Vieira de Camargo, Vedilaine Aparecida Bueno da Silva Macedo, Edvaldo Vieira de Campos, João Manoel da Silva Júnior, Nilo José Coelho Duarte, Silvia Regina Cavani Jorge Santos, David de Souza Gomez. PhD Student Ronaldo Morales Jr, Advisor Silvia R. C. J. Santos.

Award 4: Is beta-lactams effectiveness impacted by the duration of drug infusion in critically ill septic patients. A systematic review of clinical PK-studies based on PK/PD approach. Karina B Vianna, João Manoel da Silva Júnior, Elson Mendes da Silva Júnior, David de Souza Gomez, Silvia Regina

Cavani Jorge Santos. Bachelor of Pharmaceutical Sciences Karina B. Vianna, Advisor Silvia R. C. J. Santos.

Publications at the last three years:

1) Otofuji, G., Silva Jr, J. M., Silva Junior, E. M., Gomides, A. S., Rocha, A., Gomez, D. S. Lanchote V. L. & Santos, S. R. C. J. (2021). The change in medical behavior in the prescription of antimicrobials guided by cultures, biomarkers and PK/PD approach of meropenem in a septic patient with severe burns with infection caused by amikacin-resistant p aeruginosa. *The Brazilian Journal of Infectious Diseases,* 25, 101409. https://doi.org/10.1016/j.bjid.2020.101409.
2) Matsuno, V. K., Junior, J. M. S., Gomides, A. S., Silva Filho, C. R., Santos, V. J., Rocha, A., Gomez, D. S. & Santos, S. R. C. J. (2021). Fluconazole PK/PD approach to ensure effectiveness of antifungal therapy in critically ill, major burn septic patients with systemic Candida glabrata infection (CIM 32 mg/L). *The Brazilian Journal of Infectious Diseases,* 25, 101438. https://doi.org/10.1016/j.bjid.2020.101438.
3) dos Santos, M. S., Silva Jr, J. M., Silva Junior, E. M., Gomides, A. S., Rocha A., Duarte, A. J. S., Gomez, D. S. Lanchote V. L. & Santos, S. R. C. J (2021). Effectiveness of antimicrobial therapy in severe burn septic patient with renal dysfunction and severe ocular injury in intensive therapy with vancomycin and piperacillin-tazobactan through the pharmacokinetic-pharmacodynamic approach. *The Brazilian Journal of Infectious Diseases,* 25, 101337. Maria Severina dos Santos Bachelor of Pharmaceutical Sciences. https://doi.org/10.1016/j.bjid.2020.101337.
4) Souza, V. K., Silva Jr, J. M., Junior, E. M. S., Ferreira, G. A., Silva Filho, C. R., Santos, V. J., Gomez, D. S. & Santos, S. R. C. J (2021). Comparative study of the effectiveness of piperacillin-tazobactan after intermittent versus extended infusion in severe burn septic patients by the pharmacokinetic-pharmacodynamic approach (PK/PD). *The Brazilian Journal of Infectious Diseases,* 25, 101387. https://doi.org/10.1016/j.bjid.2020.101387.
5) De Camargo, T. V., da Silva Jr, J. M. M., da Silva Junior, E. M., da Silva Filho, C. R., Santos, V. J., Oliveira, T. C., Gomez, D. S. & Santos, S. R. C. J. (2021). Comparative study of the effectiveness of meropenem in septic burn patients. Teenagers versus young adults. *The Brazilian Journal of Infectious Diseases,* 25, 101388. https://doi.org/10.1016/j.bjid.2020.101388.

6) Romano, P., Duarte, N. J. C., Silva Junior, J. M., Silva Junior, E. M., Sandre, A., Rocha, Duarte, A. J. S., A Gomez, D. S., Lanchote V. L. & Santos, S. R. C. J. (2021). Effectiveness of antimicrobial therapy in a large burnt elderly septic patient in intensive care with vancomycin-meropenem through the pharmacokinetic-pharmacodynamic approach. *The Brazilian Journal of Infectious Diseases,* 25, 101403. https://doi.org/10.1016/j.bjid.2020.101403.
7) Vianna, K. B., da Silva Jr, J. M., da Silva Jr, É. M., Oliveira, T. C., Silva Filho, C. R., Santos, V. J., Rocha A, Lanchote, V. L., Gomez, D. S. & Santos, S. R. C. J. (2021). Efficacy of meropenem after rapid versus extended infusion in septic burn patients by pharmacokinetic/pharmacodynamic approach (PK/PD). *The Brazilian Journal of Infectious Diseases,* 25, 101391. https://doi.org/10.1016/j.bjid.2020.101391.
8) Junior, R. M., Juodinis, V. D. A., de Souza, D. C., do Carmo Albuquerque, A. O., Benis, C. R. C., Oliveira, S. C., Karina Fonzar Braga & Santos, S. R. C. J. (2021). Pharmacokinetic and pharmacodynamic approach in the therapeutic monitoring of vancomycin in pediatric patients after liver transplantation: case series. *Brazilian Journal of Infectious Diseases.* 101344. https://doi.org/10.1016/j.bjid.2020.101344.
9) Souza, V. K., Silva Jr J. M., Silva Jr E. M., Ferreira, G. A., Silva Filho, C. R., Santos, V. J., Rocha A., Lanchote, V. L., Gomez D. S., Santos S. R. C. J. Comparative study of the effectiveness of piperacillin-tazobactan after intermittent versus extended infusion in severe burn septic patients by the pharmacokinetic-pharmacodynamic approach (PK/PD) *The Brazilian Journal of Infectious Diseases,* 26,101337. https://doi.org/10.1016/j.bjid.2021.101337.
10) Otofuji, G., Morales Junior, R, Silva Junior, J. M., Silva Junior E. M., Oliveira, A. M. R. R., Oliveira, E. M., Ferreira G. A., Oliveira T. C., Gomez, D. S & Santos, S. R. (2022). PK/PD approach on the effectiveness of piperacillin-vancomycin in septic burn patients on an empiric regimen of the recommended dose in renal failure. *The Brazilian Journal of Infectious Diseases,* 26, 101737. https://doi.org/10.1016/j.bjid.2021.101737.
11) De Camargo, T. C., Kupa, L. V. K., Silva Junior, J. M., Silva Junior, E. M, Oliveira, A. M. R. R., Oliveira, E. M., Gomides A. S., Ferreira G. A., Oliveira T. C., Gomez, D. S. & Santos, S. R. (2022Alteration of clinical management in critically ill septic patients with burns by adjusting the dose of vancomycin against gram-positive strains CIM 2 mg/L by the

pharmacokinetic-pharmacodynamic approach. *Braz. Journal of Infectious Diseases*, 26, 101738. https://doi.org/10.1016/j.bjid.2021.101738.

12) De Camargo, T. V., da Silva Jr, J. M. M., da Silva Junior, E. M., da Silva Filho, C. R., Santos, V. J., Oliveira, T. C., Gomez, D. S. & Santos, S. R. C. J. Comparative study of the effectiveness of meropenem in septic burn patients. Teenagers versus young adults. *The Brazilian Journal of Infectious Diseases,* 25, 101388 (2021). https://doi.org/10.1016/j.bjid.2020.101388.

13) Morales Junior, R. M., Kupa, L. D., V. K., de Oliveira, E. M., de Campos, E. V., da Silva Junior, J. M., Junior, E. M. S., Gomez, D. S. & Santos, S. R. C. J. (2022). Antimicrobial coverage of vancomycin against gram-positive pathogens in burn patients. *The Brazilian Journal of Infectious Diseases,* 26, 101745. https://doi.org/10.1016/j.bjid.2021.101745.

14) Messiano, C. G., Silva Junior, E. M., Silva Junior, J. M., Gomides, A. S., Ferreira, G. A., de Oliveira, T. C., David S. Gomez, & Santos, S. R. (2022). Comparison of the effectiveness of meropenem in the early versus late phases of septic shock using the pharmacokinetic-pharmacodynamic approach in burn patients. *The Brazilian Journal of Infectious Diseases,* 26, 101746. https://doi.org/10.1016/j.bjid.2021.101746.

15) Junior, R. M., Tiguman, G. M. B., Juodinis, V. D. A., de Freitas Santos, I. C. P., & Santos, S. R. C. J. (2022). Therapeutic monitoring of vancomycin: how we are in Brazil ?. *The Brazilian Journal of Infectious Diseases,* 26, 101707. https://doi.org/10.1016/j.bjid.2021.101707.

16) Romano, P., Ebner, P. A., dos Santos, M. S., Garcia, M. S., Kupa, L. D. V. K., Duarte, N. J. C., Gomez, D. S. & Santos, S. R. (2022). Piperacillin-meropenem serum monitoring by LCMS/MS to evaluate the effectiveness of the empirical dose regimen and the extended infusion strategy in septic burn patients through the pharmacokinetic-pharmacodynamic approach. *The Brazilian Journal of Infectious Diseases,* 26, 101756. https://doi.org/10.1016/j.bjid.2021.101756.

17) De Oliveira, E. M., de Almeida, J. P., Silva Junior, J. M., de Camargo, T. V., Sabanai, A. H., Siqueira, R. S., Gomez, D. S. & Santos, S. R. C. J. (2022). Optimization of vancomycin therapy in critically ill oncology patients in intensive care against gram-positive patients up to MIC 2 mg/L based on the pharmacokinetic-pharmacodynamic approach (PK/PD). *The Brazilian Journal of Infectious Diseases,* 26, 101759. https://doi.org/10.1016/j.bjid.2021.101759.

18) Morales, R. J., Leite, F., Juodinis, V., Santos, I., Okumura, P., Ribeiro, B., & Santos, S. (2021). Vancomycin area under the curve–guided monitoring

in critically ill patients. *International Journal of Antimicrobial Agents,* 58, 21002752. https://doi.org/10.1016/j.ijantimicag.2021.106421.4.

19) Morales, Jr., R., Kupa, L. V. K., Santos, V. J., Romano, P., Oliveira, E. M., Campos, E. V, Silva Jr, J. M., Silva Jr, E. M., Gomides A. S., Gomez, D. S. & Santos, S. R. C. J. (2021). Vancomycin dose adjustment against Gram-positive MIC 2 mg/L strains in critically ill adult burn patients by pharmacokinetic-pharmacodynamic approach. In *International Journal of Antimicrobial Agents.* Elsevier. https://doi.org/10.1016/j.ijantimicag.2021.106421.8.
20) Morales, Jr, R., Kupa, L. V. K., Silva Jr, J. M., Silva Jr, E. M., Oliveira, A. M. R. R., Gomides, A. S., Ferreira G. A., Gomez, D. S. & Santos, S. R. C. J. (2021). Meropenem extended infusion against nosocomial MIC 4 mg/L strains to guarantee drug effectiveness in burn patients at the earlier period of septic shock. *International Journal of Antimicrobial Agents,* 58, 21002806. https://doi.org/10.1016/j.ijantimicag.2021.106421.9.
21) Morales, Jr, R., Kupa, L. V. K., Vianna, K. B., Garcia, M. S., Santos, V. J., Campos, E. V., Silva Jr, J. M., Silva Jr E. M., Oliveira, T. C., Gomez, D. S. & Santos, S. R. C. J. (2021). Meropenem extended infusion versus intermittent infusion against nosocomial MIC 4 mg/L strains to guarantee drug effectiveness by PK/PD approach in burn patients at the earlier period of septic shock. In *International Journal of Antimicrobial Agents.* Elsevier. https://doi.org/10.1016/j.ijantimicag.2021.106421.10.
22) Morales, Jr, R., Macedo, V. A. S., Pires, F. R., Campos, E. V., Ebner, P. A. R., Romano, P., Maria Severina dos Santos, Duarte, N. J., Gomez, D. S. & Santos, S. R. C. J. Vancomycin target attainment in critically ill septic pediatric burn versus non-burn patients against Staphylococcus spp. *International Journal of Antimicrobial Agents,* 58, 2100282. https://doi.org/10.1016/j.ijantimicag.2021.106421.11.
23) Morales, Jr, R., Chalom, M. Y., Romano, P., Ebner, P. A. R., Santos, M. S., Kupa, L. V. K., Sumita, N. M., Duarte N. J. C., Duarte, A. J. S., Gomez, D. S. & Santos, S. R. J. C. (2021). Meropenem 3 hrs-extended infusion improves effectiveness in critically ill burn patients at the earlier period of septic shock against P. aeruginosa intermediate susceptibility MIC 4-8 mg/L In *International Journal of Antimicrobial Agents.* Elsevier. https://doi.org/10.1016/j.ijantimicag.2021.106421.12.
24) Morales, Jr, R., Chalom, M. Y., Romano, P., Ebner, P. A. R., Santos, M. S., Kupa, L. V. K., Sumita, N. M., Duarte N. J. C., Duarte, A. J. S., Gomez, D. S. & Santos, S. R. J. C. (2021). Piperacillin/tazobactam effectiveness in septic burn patients requiring vasopressors by applying

pharmacokinetics-pharmacodynamics approach. In *International Journal of Antimicrobial Agents.* Elsevier. https://doi.org/10.1016/j.ijantimicag.2021.106421.13.

25) Morales, Jr, R., Chalom, M. Y., Romano, P., Ebner, P. A. R., Santos, M. S., Garcia, M. S., Kupa, L. V. K., Duarte N. J. C., Duarte, A. J. S., Gomez, D. S. & Santos, S. R. J. C. (2021). Extended infusion improves meropenem and piperacillin/tazobactam effectiveness in septic burn patients with normal renal by applying pharmacokinetics-pharmacodynamics approach. In *International Journal of Antimicrobial Agents.* Elsevier. https://doi.org/10.1016/j.ijantimicag.2021.106421.14.
26) Souza, V. K., Silva Jr E. M., Kaneko V. M., Silva Jr J. M., Gomides A. S., Ferreira G. A., Oliveira T. C., Gomez D. S., Santos, S. R. C. J. Extended two hours infusion improves piperacillin effectiveness against susceptible strains in septic burn patients. *Critical Care* (2019). v.23(3): p 21-22, https://doi.org/10.1186/s13054-019-9. ISSN 1364-8535.
27) Messiano, C. G., Junior, R. M., Pereira, G. O., da Silva Junior, E. M., Gomez, D. S. & Santos, S. R. C. J. (2022). Therapeutic Target Attainment of 3-Hour Extended Infusion of Meropenem in Patients with Septic Burns. *Clinical Therapeutics*, 44(4), 624-629. http://dx.doi.org/10.1016/j.clinthera.2022.02.007.
28) De Camargo, T. V., Junior, E. M. S., Silva Jr, J. M., Gomez, D. S. & Santos, S. R. C. J. (2022). PK/PD approach to evaluate Meropenem effectiveness in critically ill burn adolescents versus young adults undergoing therapy of septic shock. *Pharm Pharmacol Int J*, 10(3), 79-85 (2022). https://doi.org/10.15406/ppij.2022.10.00368.
29) Santos, S. R. C. J., de Camargo T. V., Macedo V. A., Souza, V. K. & Gomez, D. S. Vancomycin and meropenem serum monitoring for target attainment by PK/PD approach as an effective tool in the battle against nosocomial pathogens in septic pediatric burn patients. *Pharm Pharmacol Int J.* 2022;10(3):87–95 (2022). https://doi.org/10.15406/ppij.2022.10.00369.
30) Santos, S. R. C. J., de Camargo T. V., Messiano C. G., Kupa L. V. K., Souza V. K., Ronaldo Morales Jr. R., Sanches Pinto D. C. S., Silva Junior E. M., Silva Junior J. M., Gomez D. S. Combating bacterial resistance to antimicrobials in severe septic ICU patients: importance of meropenem, piperacillin serum monitoring as a dose adjustment and duration of infusion strategies. *Pharm Pharmacol Int J.* 11(2):52–60 (2023a). https://doi.org/10.15406/ppij.2023.11.00402.

31) Santos, S. R. C. J., de Camargo T. V., Messiano C. G., Souza V. K. Frederico Ribeiro Pires, Débora Cristina Sanches Pinto, Edvaldo Vieira de Campos, Elson Mendes da Silva Junior, João Manoel da Silva Junior, David de Souza Gomez. Combined vancomycin-meropenem therapy in pediatric major burns undergoing therapy of septic shock guided by cultures and pharmacokinetic-pharmacodynamics approach based on serum levels to combat bacterial resistance. *Pharm Pharmacol Int J.* 11(2):62–68 (2023b). https://doi.org/10.15406/ppij.2023.11.00403.

Claudia Garcia Messiano

ORCID https://orcid.org/0000-0002-6516-4909. Brazilian Platform of Research (CNPq).

Affiliation:
University of São, School of Pharmaceutical Sciences - Master of Pharmacy (2021).

Education:
University of Oswaldo Cruz: Pharmacist-Biochemistry; Master of Pharmaceutical Sciences (2021); Research fellow of Clinical Pharmacokinetics Team.

Research Experience:
Burn patients in hospital – Medical School of University of São Paulo (2019-2023).

Professional Experience:
Antimicrobial effectiveness and safety by individualization of therapy in ICU major burn patients by therapeutic drug monitoring and PK/PD approach based on serum levels at the Clinical Pharmacokinetics Center – University of Sao Paulo.

Professional Appointments:
Clinical Pharmacist in Health Sciences in public hospital.

Honors:

Member of Research-Team from Clinical Pharmacokinetics Group since 2022.

Awards at the Last three years

III International Meeting Sobratafe on Advanced Wound Care, Sobratfe 2019, Maksoud Plaza Hotel São Paulo, SP. Awards At The Event Promoter: Brazilian Society for Advanced Treatment of Complex Wounds (SOBRATAFE) "Clinical Protocol"

Award 1: Meropenem effectiveness in critically ill burn patients during the septic shock by pharmacokinetics assessment based on drug serum measurements done in a real time. Claudia Garcia Messiano, Elson Mendes da Silva Júnior, João Manoel da Silva Júnior, Edvaldo Vieira de Campos, Amanda Maria Ribas Rosa de Oliveira, Aline S. Gomides, Gabriela A. Ferreira, Thiago C. Oliveira, Silvia Regina Cavani Jorge Santos, David de Souza Gomez.

Research team member since 2018

Publications from the Last 3 Years:

1) Messiano, C. G., Silva Junior, E. M., Silva Junior, J. M., Gomides, A. S., Ferreira, G. A., de Oliveira, T. C., David S. Gomez, & Santos, S. R. (2022). comparação da efetividade do meropenem nas fases precoce versus tardia do choque séptico através da abordagem farmacocinética-farmacodinâmica em pacientes queimados. *The Brazilian Journal of Infectious Diseases,* 26, 101746. https://doi.org/10.1016/j.bjid.2021.101746.
2) Messiano, C. G., Junior, R. M., Pereira, G. O., da Silva Junior, E. M., Gomez, D. S. & Santos, S. R. C. J. (2022). Therapeutic Target Attainment of Extended 3 hrs.-Infusion of Meropenem in Patients with Septic Burns. *Clinical Therapeutics,* 44(4), 624-629. http://dx.doi.org/10.1016/j.clinthera.2022.02.007.
3) Santos, S. R. C. J., de Camargo T. V., Messiano C. G., Kupa L. V. K., Souza V. K., Ronaldo Morales Jr. R., Sanches Pinto D. C. S., Silva Junior E. M., Silva Junior J. M., Gomez D. S. Combating bacterial resistance to antimicrobials in severe septic ICU patients: importance of meropenem, piperacillin serum monitoring as a dose adjustment and duration of

infusion strategies. *Pharm Pharmacol Int J.* 2023;11(2):52–60 (2023). https://doi.org/10.15406/ppij.2023.11.00403.

4) Santos, S. R. C. J., de Camargo T. V., Messiano C. G., Souza V. K. Frederico Ribeiro Pires, Débora Cristina Sanches Pinto, Edvaldo Vieira de Campos, Elson Mendes da Silva Junior, João Manoel da Silva Junior, David de Souza Gomez. Combined vancomycin-meropenem therapy in pediatric major burns undergoing therapy of septic shock guided by cultures and pharmacokinetic-pharmacodynamics approach based on serum levels to combat bacterial resistance. *Pharm Pharmacol Int J.* 2023;11(2):62–68 (2023). https://doi.org/10.15406/ppij.2023.11.00403.

Thais Vieira de Camargo

ORCID https://orcid.org/0000-0002-4799-7122. Brazilian Platform of Research (CNPq).

Affiliation:
University of São Paulo, School of Pharmaceutical Sciences - Master of Pharmacy (2021).

Education:
University of Piracicaba Methodist: Pharmacist; Master of Pharmaceutical Sciences (2021); PhD student of Medical School of University of Sao Paulo. Pharmacist in Health Sciences of Federal Health Institute, Sao Paulo/SP Brazil.

Research Experience:
Burn patients in hospital – Medical School of University of São Paulo (2019-2023).

Professional Experience:
Antimicrobial effectiveness and safety by individualization of therapy in ICU major burn patients by therapeutic drug monitoring and PK/PD approach based on serum levels at the Clinical Pharmacokinetics Center – University of Sao Paulo.

Professional Appointments:

Teaches at the High School "Paula Souza Center"; Inspector of State Health Surveillance Center (Centro de Vigilance Sanitaria Estadual, CVS) of the Federal Health Surveillance Center, Sao Paulo/SP Brazil.

Honors: Member of Research-Team from Clinical Pharmacokinetics Group since 2018.

Scientific Awards from the Last 3 Years:

National Meeting Sobratafe on Advanced Wound Care, Sobratfe 2020, Maksoud Plaza Hotel São Paulo, SP. FOUR Awards at The Event. Promoter: Brazilian Society for Advanced Treatment of Complex Wounds (Sobratafe) *"Clinical Protocol" Awards Dr Walter Soares Pinto Prize*

Award 1: Is Meropenem effectiveness in ICU burn patients during SIRS at the earlier period of septic shock impacted by pharmacokinetic changes? Adolescents versus Young adults. Thais V. de Camargo, Edvaldo Vieira de Campos, Amanda Maria Ribas Rosa de Oliveira, João Manoel da Silva Júnior, Elson Mendes da Silva Júnior, Aline S. Gomides, Gabriela A. Ferreira, Thiago C. Oliveira, Vera Lúcia Lanchote, Silvia Regina Cavani Jorge Santos, David de Souza Gomez.

Research team member since 2017

Publications from the Last 3 Years:

1) De Camargo, T. V., da Silva Jr, J. M. M., da Silva Junior, E. M., da Silva Filho, C. R., Santos, V. J., Oliveira, T. C., Gomez, D. S. & Santos, S. R. C. J. (2021). Estudo comparativo da efetividade do meropenem em pacientes sépticos queimados. adolescentes versus adultos jovens. *The Brazilian Journal of Infectious Diseases,* 25, 101388. https://doi.org/10.1016/j.bjid.2020.101388.
2) De Camargo T. C., Kupa, L. V. K., Silva Junior, J. M., Silva Junior, E. M, Oliveira, A. M. R. R., Oliveira, E. M., Gomides A. S., Ferreira G. A., Oliveira T. C., Gomez, D. S. & Santos, S. R. (2022). Alteração de conduta clínica em pacientes sépticos gravemente enfermos com queimaduras pelo ajuste da dose da vancomicina contra cepas gram-positivas CIM 2 mg/l pela abordagem farmacocinética-farmacodinâmica. *The Brazilian Journal of Infectious Diseases*, 26, 101738. https://doi.org/10.1016/j.bjid.2021.101738.

3) De Camargo, T. V., da Silva Jr, J. M. M., da Silva Junior, E. M., da Silva Filho, C. R., Santos, V. J., Oliveira, T. C., Gomez, D. S. & Santos, S. R. C. J. (2021). Estudo comparativo da efetividade do meropenem em pacientes sépticos queimados. adolescentes versus adultos jovens. *The Brazilian Journal of Infectious Diseases,* 25, 101388. https://doi.org/10.1016/j.bjid.2020.101388.
4) De Oliveira, E. M., de Almeida, J. P., Silva Junior, J. M., de Camargo, T. V., Sabanai, A. H., Siqueira, R. S., Gomez, D. S. & Santos, S. R. C. J. (2022). Otimização da terapia com a vancomicina nos pacientes críticos oncológicos em terapia intensiva contra gram-positivos até CIM 2 mg/l baseada na abordagem farmacocinética-farmacodinâmica (PK/PD). *The Brazilian Journal of Infectious Diseases,* 26, 101759. https://doi.org/10.1016/j.bjid.2021.101759.
5) De Camargo, T. V., Junior, E. M. S., Silva Jr, J. M., Gomez, D. S. & Santos, S. R. C. J. (2022). PK/PD approach to evaluate Meropenem effectiveness in critically ill burn adolescents versus young adults undergoing therapy of septic shock. *Pharm Pharmacol Int J*, 10(3), 79-85 (2022). https://doi.org/10.15406/ppij.2022.10.00368.
6) Santos, S. R. C. J., de Camargo T. V., Macedo V. A., Souza, V. K. & Gomez, D. S. Vancomycin and meropenem serum monitoring for target attainment by PK/PD approach as an effective tool in the battle against nosocomial pathogens in septic pediatric burn patients. *Pharm Pharmacol Int J.* 2022;10(3):87–95 (2022). https://doi.org/10.15406/ppij.2022.10.00369.
7) Santos, S. R. C. J., de Camargo T. V., Messiano C. G., Kupa L. V. K., Souza V. K., Ronaldo Morales Jr. R., Sanches Pinto D. C. S., Silva Junior E. M., Silva Junior J. M., Gomez D. S. Combating bacterial resistance to antimicrobials in severe septic ICU patients: importance of meropenem, piperacillin serum monitoring as a dose adjustment and duration of infusion strategies. *Pharm Pharmacol Int J.* 11(2):52–60 (2023). https://doi.org/10.15406/ppij.2023.11.00402.
8) Santos, S. R. C. J., de Camargo T. V., Messiano C. G., Souza V. K. Frederico Ribeiro Pires, Débora Cristina Sanches Pinto, Edvaldo Vieira de Campos, Elson Mendes da Silva Junior, João Manoel da Silva Junior, David de Souza Gomez. Combined vancomycin-meropenem therapy in pediatric major burns undergoing therapy of septic shock guided by cultures and pharmacokinetic-pharmacodynamics approach based on serum levels to combat bacterial resistance. *Pharm Pharmacol Int J.* 11(2):62–68 (2023). https://doi.org/10.15406/ppij.2023.11.00403.

Vanessa Kasubeck de Souza

ORCID https://orcid.org/0000-0002-6634-5867. Brazilian Platform of Research (CNPq).

Affiliation:
University of São Paulo, School of Pharmaceutical Sciences - Master of Pharmacy (2021).

Education:
University of São Paulo: Pharmacist-Biochemistry; Clinical Pharmacy Expertise, Master of Pharmaceutical Sciences (2021). Clinical Pharmacist "Health Sciences" in hospital.

Research Experience:
Burn patients in hospital – Medical School of University of São Paulo (2019-2023).

Professional Experience:
Antimicrobial effectiveness and safety by individualization of therapy in ICU major burn patients by therapeutic drug monitoring and PK/PD approach based on serum levels at the Clinical Pharmacokinetics Center – University of Sao Paulo.

Professional Appointments:
Clinical Pharmacist in Health Sciences in hospital.

Honors: Member of Research-Team from Clinical Pharmacokinetics Group since 2016.

Scientific Awards from the Last 3 Years:
Brazilian Society of Infectiology (BSID)PRIZE 2021. The Brazilian Journal of Infectious Diseases, Volume 25, Supplement 1, January 2021 ISSN 1413-86-70. https://doi.org/10.1016/j.bjid.2020.101387. "Comparative study of the effectiveness of Piperacillin-Tazobactam after intermittent versus extended infusion in severe burn septic patients by the pharmacokinetic-pharmacodynamic approach (PK/PD)," Master in Pharmaceutical Sciences Vanessa Kasubeck de Souza.

Global Meeting Sobratafe on Advanced Wound Care, Sobratfe 2020, Maksoud Plaza Hotel São Paulo, SP. FOUR Awards at The Event. Promoter: Brazilian Society for Advanced Treatment of Complex Wounds (Sobratafe) "Clinical Protocol" Awards Dr Walter Soares Pinto Prize 2 Extended infusion improves beta-lactams effectiveness in septic burn patients with normal renal function against P aeruginosa intermediate susceptibility. Vanessa Kasubeck de Souza, Karina B Vianna, João Manoel da Silva Júnior, Amanda Maria Ribas Rosa de Oliveira, Edvaldo Vieira de Campos, Elson Mendes da Silva Júnior, Aline S. Gomides, Gabriela A. Ferreira, Thiago C. Oliveira, Vera Lúcia Lanchote, Silvia Regina Cavani Jorge Santos, David de Souza Gomez.

Research team member since 2019

Publications from the Last 3 Years:

1) Souza, V. K., Silva Jr, J. M., Junior, E. M. S., Ferreira, G. A., Silva Filho, C. R., Santos, V. J., Gomez, D. S. & Santos, S. R. C. J (2021). Estudo comparativo da efetividade da piperacilina-tazobactana após infusão intermitente versus estendida em pacientes sépticos grandes queimados pela abordagem farmacocinética-farmacodinâmica (PK/PD). *The Brazilian Journal of Infectious Diseases*, 25, 101387. https://doi.org/10.1016/j.bjid.2020.101387.
2) Souza V. K., Silva Jr J. M., Silva Jr E. M., Ferreira, G. A., Silva Filho, C. R., Santos, V. J., Rocha A., Lanchote, V. L., Gomez D. S., Santos S. R. C. J. Estudo comparativo da efetividade da piperacilina-tazobactana após infusão intermitente versus estendida em pacientes sépticos grandes queimados pela abordagem farmacocinética-farmacodinâmica (PK/PD) *The Brazilian Journal of Infectious Diseases,* 26,101337. https://doi.org/10.1016/j.bjid.2021.101337.
3) Souza, V. K., Silva Jr E. M., Kaneko V. M., Silva Jr J. M., Gomides A. S., Ferreira G. A., Oliveira T. C., Gomez D. S., Santos, S. R. C. J. Extended two hours infusion improves piperacillin effectiveness against susceptible strains in septic burn patients. *Critical Care* (2019). v.23(3): p 21-22, https://doi.org/10.1186/s13054-019-9. ISSN 1364-8535.
4) Santos, S. R. C. J., de Camargo T. V., Macedo V. A., Souza, V. K. & Gomez, D. S. Vancomycin and meropenem serum monitoring for target attainment by PK/PD approach as an effective tool in the battle against nosocomial pathogens in septic pediatric burn patients. *Pharm Pharmacol*

Int J. 2022;10(3):87–95 (2022). https://doi.org/10.15406/ppij.2022.10.00369.

5) Santos, S. R. C. J., de Camargo T. V., Messiano C. G., Kupa L. V. K., Souza V. K., Ronaldo Morales Jr. R., Sanches Pinto D. C. S., Silva Junior E. M., Silva Junior J. M., Gomez D. S. Combating bacterial resistance to antimicrobials in severe septic ICU patients: importance of meropenem, piperacillin serum monitoring as a dose adjustment and duration of infusion strategies. *Pharm Pharmacol Int J.* 11(2):52–60 (2023). https://doi.org/10.15406/ppij.2023.11.00402.
6) Santos, S. R. C. J., de Camargo T. V., Messiano C. G., Souza V. K. Frederico Ribeiro Pires, Débora Cristina Sanches Pinto, Edvaldo Vieira de Campos, Elson Mendes da Silva Junior, João Manoel da Silva Junior, David de Souza Gomez. Combined vancomycin-meropenem therapy in pediatric major burns undergoing therapy of septic shock guided by cultures and pharmacokinetic-pharmacodynamics approach based on serum levels to combat bacterial resistance. *Pharm Pharmacol Int J.* 11(2):62–68 (2023). https://doi.org/10.15406/ppij.2023.11.00403.

Ronaldo Morales Junior

ORCID https://orcid.org/0000-0002-5856-3717. Brazilian Platform of Research (CNPq).

Affiliation:
University of São Paulo/School of Pharmaceutical Sciences, Clinical Pharmacy Expertise - Doctor of Pharmacy (2023).

Education:
University of São Paulo: Pharmacist-Biochemist; Residence two years in hospital, Sirio- Lebanese hospital, Doctor of Pharmaceutical Sciences (PhD, June 22nd, 2023). Clinical Pharmacist in Health Sciences in hospital. Post doctoral studies in Clinical Pharmacokinetics, University of Michigan USA (2023-2024).

Research Experience:
Pediatric patients in hospital, ICU of Burns– Medical School of University of São Paulo and ICU pediatric patients with liver transplantation (2020-2023).

Professional Experience: Antimicrobial effectiveness and safety by individualization of therapy in ICU major burn patients by therapeutic drug monitoring and PK/PD approach based on serum levels at the Clinical Pharmacokinetics Center – University of Sao Paulo.

Professional Appointments: Clinical Pharmacist in Health Sciences in Sirio Lebanese, private hospital.

Honors: Member of Research-Team from Clinical Pharmacokinetics Group since 2018.

Scientific Awards from the Last 3 Years:

III International Meeting Sobratafe on Advanced Wound Care, 2019.

Promoter: Brazilian Society for Advanced Treatment of Complex Wounds (Sobratafe) "Clinical Protocol" Awards Sobratfe 2019

Award 2: Vancomycin dose adjustment done in areal time for target attainment in critically ill septic pediatric burns versus non-burns against Staphylococcus spp MIC 1 mg/L. Camila Salinas Guillaux, Ronaldo Morales Júnior, Vedilaine Aparecida Bueno da Silva Macedo, Elson Mendes da Silva Júnior, João Manoel da Silva Júnior, Edvaldo Vieira de Campos, Amanda Maria Ribas Rosa de Oliveira, Aline Sandré Gomides, Gabriela Aparecida Ferreira, Thiago C. Oliveira, Silvia Regina Cavani Jorge Santos, David de Souza Gomez.

National Meeting Sobratafe on Advanced Wound Care, Sobratfe 2020. Promoter: Brazilian Society for Advanced Treatment of Complex Wounds (Sobratafe)

"Clinical Protocol Awards" Dr Walter Soares Pinto Prize Pharmacokinetic-pharmacodynamic approach permits vancomycin dose adjustment for effectiveness against infection caused by Gram-positive pathogens MIC > 1 mg/L in ICU pediatric burn patients. Ronaldo Morales Júnior, Thais Vieira de Camargo, Vedilaine Aparecida Bueno da Silva Macedo, Edvaldo Vieira de Campos, João Manoel da Silva Júnior, Nilo José Coelho Duarte, Silvia Regina Cavani Jorge Santos, David de Souza Gomez.

Asian Pacific Foundation for Infectious Diseases "2021 Outstanding Poster Award" 13th International Symposium on Antimicrobial Agents and Resistance (ISAAR-2021). e-Poster Presentation Winner promoted by Asian Pacific Foundation for Infectious Diseases (APFID) September 9-10/2021. Published: International Journal of Antimicrobial Agents, *Volume 58,*

Supplement 1, September 2021, 2100279-21002818 Morales, Jr, R., Kupa, L. V. K., Vianna, K. B., Garcia, M. S., Santos, V. J., Campos, E. V., Silva Jr, J. M., Silva Jr E. M., Oliveira, T. C., Gomez, D. S. & Santos, S. R. C. J. (2021). Meropenem extended infusion versus intermittent infusion against nosocomial MIC 4 mg/L strains to guarantee drug effectiveness by PK/PD approach in burn patients at the earlier period of septic shock. In *International Journal of Antimicrobial Agents*. Elsevier. https://doi.org/10.1016/j.ijantimicag.2021.106421.10.

Research team member since 2018

Publications from the Last 3 Years:

1) Junior, R. M., Juodinis, V. D. A., de Souza, D. C., do Carmo Albuquerque, A. O., Benis, C. R. C., Oliveira, S. C., Karina Fonzar Braga & Santos, S. R. C. J. (2021). Abordagem farmacocinética e farmacodinâmica no monitoramento terapêutico de vancomicina em pacientes pediátricos pós-transplante hepático: série de casos. *Brazilian Journal of Infectious Diseases*. 101344. https://doi.org/10.1016/j.bjid.2020.101344.
2) Otofuji, G., Morales Junior, R, Silva Junior, J. M., Silva Junior E. M., Oliveira, A. M. R. R., Oliveira, E. M., Ferreira G. A., Oliveira T. C., Gomez, D. S & Santos, S. R. (2022). Abordagem PK/PD na efetividade de piperacilina-vancomicina em pacientes sépticos queimados em regime empírico da dose recomendada na insuficiência renal. *The Brazilian Journal of Infectious Diseases,* 26, 101737. https://doi.org/10.1016/j.bjid.2021.101737.
3) Morales Junior, R. M., Kupa, L. D., V. K., de Oliveira, E. M., de Campos, E. V., da Silva Junior, J. M., Junior, E. M. S., Gomez, D. S. & Santos, S. R. C. J. (2022). Cobertura antimicrobiana da vancomicina contra patógenos gram-positivos em pacientes queimados. *The Brazilian Journal of Infectious Diseases,* 26, 101745. https://doi.org/10.1016/j.bjid.2021.101745.
4) Junior, R. M., Tiguman, G. M. B., Juodinis, V. D. A., de Freitas Santos, I. C. P., & Santos, S. R. C. J. (2022). Monitoramento terapêutico de vancomicina: como estamos no Brasil?. *The Brazilian Journal of Infectious Diseases,* 26, 101707. https://doi.org/10.1016/j.bjid.2021.101707.
5) Morales, R. J., Leite, F., Juodinis, V., Santos, I., Okumura, P., Ribeiro, B., & Santos, S. (2021). Vancomycin area under the curve–guided monitoring

in critically ill patients. *International Journal of Antimicrobial Agents,* 58, 21002752. https://doi.org/10.1016/j.ijantimicag.2021.106421.4.

6) Morales, Jr., R., Kupa, L. V. K., Santos, V. J., Romano, P., Oliveira, E. M., Campos, E. V, Silva Jr, J. M., Silva Jr, E. M., Gomides A. S., Gomez, D. S. & Santos, S. R. C. J. (2021). Vancomycin dose adjustment against Gram-positive MIC 2 mg/L strains in critically ill adult burn patients by pharmacokinetic-pharmacodynamic approach. In *International Journal of Antimicrobial Agents.* Elsevier. https://doi.org/10.1016/j.ijantimicag.2021.106421.8.
7) Morales, Jr, R., Kupa, L. V. K., Silva Jr, J. M., Silva Jr, E. M., Oliveira, A. M. R. R., Gomides, A. S., Ferreira G. A., Gomez, D. S. & Santos, S. R. C. J. (2021). Meropenem extended infusion against nosocomial MIC 4 mg/L strains to guarantee drug effectiveness in burn patients at the earlier period of septic shock. *International Journal of Antimicrobial Agents,* 58, 21002806. https://doi.org/10.1016/j.ijantimicag.2021.106421.9.
8) Morales, Jr, R., Kupa, L. V. K., Vianna, K. B., Garcia, M. S., Santos, V. J., Campos, E. V., Silva Jr, J. M., Silva Jr E. M., Oliveira, T. C., Gomez, D. S. & Santos, S. R. C. J. (2021). Meropenem extended infusion versus intermittent infusion against nosocomial MIC 4 mg/L strains to guarantee drug effectiveness by PK/PD approach in burn patients at the earlier period of septic shock. In *International Journal of Antimicrobial Agents*. Elsevier. https://doi.org/10.1016/j.ijantimicag.2021.106421.10.
9) Morales, Jr, R., Macedo, V. A. S., Pires, F. R., Campos, E. V., Ebner, P. A. R., Romano, P., Maria Severina dos Santos, Duarte, N. J., Gomez, D. S. & Santos, S. R. C. J. Vancomycin target attainment in critically ill septic pediatric burn versus non-burn patients against Staphylococcus spp. *International Journal of Antimicrobial Agents*, 58, 2100282. https://doi.org/10.1016/j.ijantimicag.2021.106421.11.
10) Morales, Jr, R., Chalom, M. Y., Romano, P., Ebner, P. A. R., Santos, M. S., Kupa, L. V. K., Sumita, N. M., Duarte N. J. C., Duarte, A. J. S., Gomez, D. S. & Santos, S. R. J. C. (2021). Meropenem 3 hrs-extended infusion improves effectiveness in critically ill burn patients at the earlier period of septic shock against P. aeruginosa intermediate susceptibility MIC 4-8 mg/L In *International Journal of Antimicrobial Agents*. Elsevier. https://doi.org/10.1016/j.ijantimicag.2021.106421.12.
11) Morales, Jr, R., Chalom, M. Y., Romano, P., Ebner, P. A. R., Santos, M. S., Kupa, L. V. K., Sumita, N. M., Duarte N. J. C., Duarte, A. J. S., Gomez, D. S. & Santos, S. R. J. C. (2021). Piperacillin/tazobactam effectiveness in septic burn patients requiring vasopressors by applying

pharmacokinetics-pharmacodynamics approach. In *International Journal of Antimicrobial Agents.* Elsevier. https://doi.org/10.1016/j.ijantimicag.2021.106421.13.

12) Morales, Jr, R., Chalom, M. Y., Romano, P., Ebner, P. A. R., Santos, M. S., Garcia, M. S., Kupa, L. V. K., Duarte N. J. C., Duarte, A. J. S., Gomez, D. S. & Santos, S. R. J. C. (2021). Extended infusion improves meropenem and piperacillin/tazobactam effectiveness in septic burn patients with normal renal by applying pharmacokinetics-pharmacodynamics approach. In *International Journal of Antimicrobial Agents.* Elsevier. https://doi.org/10.1016/j.ijantimicag.2021.106421.14.
13) Morales Junior, R., Telles J. P., Kwiatkowski S. Y. C., Juodinis V. D., de Souza D. C., Santos S. R. C. J. Pharmacokinetic and pharmacodynamic considerations of antibiotics and antifungals in liver transplantation recipients. *Liver Transpl* (2022) https://doi.org/10.1002/lt.26517.
14) Morales Junior, R., Telles, J. P., Kwiatkowski, S. Y. C., Juodinis, V. D. A., de Souza, D. C., & Santos, S. R. C. J. Pharmacokinetic and pharmacodynamic considerations of antibiotics and antifungals in liver transplantation recipients. *Liver Transplantation* (2023). doi: https://doi.org/10.1002/lt.26517.
15) Morales Junior R., Pereira G. O., Tiguman G. M. B., Juodinis V. D., Telles J. P., de Souza D. C., Santos S. R. C. J. Beta-Lactams Therapeutic Monitoring in Septic Children-What Target Are We Aiming for? A Scoping Review. *Front Pediatr* 10:777854 (2022). https://doi.org/10.3389/fped.2022.777854.
16) Morales Junior R., Juodinis V. D., Santos I. C.P de F., Campioni C. C., Wirth F. G. de A., Barbosa L. M. G., Souza D. C. de, Santos S. R. C. J. Vancomycin area under the curve-guided monitoring in pediatric patients. *Rev bras ter intensiva* 34:147–153 (2022). https://doi.org/10.5935/0103-507X.20220009.
17) Messiano, C. G., Junior, R. M., Pereira, G. O., da Silva Junior, E. M., Gomez, D. S. & Santos, S. R. C. J. (2022). Therapeutic Target Attainment of 3-Hour Extended Infusion of Meropenem in Patients with Septic Burns. *Clinical Therapeutics*, 44(4), 624-629. http://dx.doi.org/10.1016/j.clinthera.2022.02.007.
18) Santos, S. R. C. J., de Camargo T. V., Messiano C. G., Kupa L. V. K., Souza V. K., Ronaldo Morales Jr. R., Sanches Pinto D. C. S., Silva Junior E. M., Silva Junior J. M., Gomez D. S. Combating bacterial resistance to antimicrobials in severe septic ICU patients: importance of meropenem, piperacillin serum monitoring as a dose adjustment and duration of

infusion strategies. *Pharm Pharmacol Int J.* 11(2):52–60 (2023). https://doi.org/10.15406/ppij.2023.11.00402.
19) Santos, S. R. C. J., de Camargo T. V., Messiano C. G., Souza V. K. Frederico Ribeiro Pires, Débora Cristina Sanches Pinto, Edvaldo Vieira de Campos, Elson Mendes da Silva Junior, João Manoel da Silva Junior, David de Souza Gomez. Combined vancomycin-meropenem therapy in pediatric major burns undergoing therapy of septic shock guided by cultures and pharmacokinetic-pharmacodynamics approach based on serum levels to combat bacterial resistance. *Pharm Pharmacol Int J.* 11(2):62–68 (2023). https://doi.org/10.15406/ppij.2023.11.00403.

Leonard de Vinci Kanda Kupa

ORCID https://orcid.org/0000-0002-8827-1360. Brazilian Platform of Research (CNPq).

Affiliation:

University of São Paulo/School of Pharmaceutical Sciences, Clinical Pharmacy Expertise, Master of Toxicology (2016), PhD. Doctor of Pharmacy/Expertise Toxicology- Clinical Pharmacokinetics (2019).

Education:

University of São Paulo: Pharmacist-Biochemistry; Doctor of Pharmaceutical Sciences (2019). Clinical Pharmacist in Health Sciences in hospital. Research fellow Post doctoral studies – Inflammatory diseases in Clinical Medicine Department of Medical School, University of Sao Paulo since 2019-2023.

Research Experience:

Patients in hospital ICU of Burns, Systemic Lupus erythematosus patients – Medical School of University of São Paulo (2019-2023).

Professional Experience:

Antimicrobial effectiveness and safety by individualization of therapy in ICU major burn patients by therapeutic drug monitoring and PK/PD approach based on serum levels at the Clinical Pharmacokinetics Center – University of Sao Paulo. Vaccines against SARS CoV-2 studies in Systemic Lupus erythematosus outpatients of tertiary public hospital (HC FM USP).

Professional Appointments: Research fellow and Clinical Pharmacist in Health Sciences in our hospital.

Honors: Member of Research-Team from Clinical Pharmacokinetics Group since 2015.

Scientific Awards from the Last 3 Years:

Asian Pacific Foundation for Infectious Diseases "2021 Outstanding Poster Award" 13th International Symposium on Antimicrobial Agents and Resistance (ISAAR-2021).

Poster Presentation Winner promoted by Asian Pacific Foundation for Infectious Diseases (APFID) September 9-10/2021. Published: International Journal of Antimicrobial Agents, *Volume 58, Supplement 1*, September 2021, 2100279-21002818 R. Morales, Jr, R., Kupa, L. V. K., Vianna, K. B., Garcia, M. S., Santos, V. J., Campos, E. V., Silva Jr, J. M., Silva Jr E. M., Oliveira, T. C., Gomez, D. S. & Santos, S. R. C. J. (2021). Meropenem extended infusion versus intermittent infusion against nosocomial MIC 4 mg/L strains to guarantee drug effectiveness by PK/PD approach in burn patients at the earlier period of septic shock. In *International Journal of Antimicrobial Agents.* Elsevier. https://doi.org/10.1016/j.ijantimicag.2021.106421.10.

Research team member since 2015

Publications from the Last 3 Years:

1) de Camargo T. C., Kupa, L. V. K., Silva Junior, J. M., Silva Junior, E. M, Oliveira, A. M. R. R., Oliveira, E. M., Gomides A. S., Ferreira G. A., Oliveira T. C., Gomez, D. S. & Santos, S. R. (2022). Alteração de conduta clínica em pacientes sépticos gravemente enfermos com queimaduras pelo ajuste da dose da vancomicina contra cepas gram-positivas CIM 2 mg/l pela abordagem farmacocinética-farmacodinâmica. *The Brazilian Journal of Infectious Diseases*, 26, 101738. https://doi.org/10.1016/j.bjid.2021.101738.
2) Morales Junior, R. M., Kupa, L. D., V. K., de Oliveira, E. M., de Campos, E. V., da Silva Junior, J. M., Junior, E. M. S., Gomez, D. S. & Santos, S. R. C. J. (2022). Cobertura antimicrobiana da vancomicina contra patógenos gram-positivos em pacientes queimados. *The Brazilian Journal of Infectious Diseases,* 26, 101745. https://doi.org/10.1016/j.bjid.2021.101745.

3) Romano, P., Ebner, P. A., dos Santos, M. S., Garcia, M. S., Kupa L. V. K., Duarte, N. J. C., Gomez, D. S & Santos, S. R. (2022). Monitoramento sérico de piperacilina-meropenem por LCMS/MS para avaliação da efetividade do regime de dose empírica e a estratégia de infusão estendida em pacientes sépticos queimados através da abordagem farmacocinética-farmacodinâmica. *The Brazilian Journal of Infectious Diseases*, 26, 101756. https://doi.org/10.1016/j.bjid.2021.101756.
4) Morales, Jr., R., Kupa, L. V. K., Santos, V. J., Romano, P., Oliveira, E. M., Campos, E. V, Silva Jr, J. M., Silva Jr, E. M., Gomides A. S., Gomez, D. S. & Santos, S. R. C. J. (2021). Vancomycin dose adjustment against Gram-positive MIC 2 mg/L strains in critically ill adult burn patients by pharmacokinetic-pharmacodynamic approach. In *International Journal of Antimicrobial Agents*. Elsevier. https://doi.org/10.1016/j.ijantimicag.2021.106421.8.
5) Morales, Jr, R., Kupa, L. V. K., Silva Jr, J. M., Silva Jr, E. M., Oliveira, A. M. R. R., Gomides, A. S., Ferreira G. A., Gomez, D. S. & Santos, S. R. C. J. (2021). Meropenem extended infusion against nosocomial MIC 4 mg/L strains to guarantee drug effectiveness in burn patients at the earlier period of septic shock. *International Journal of Antimicrobial Agents*, 58, 21002806. https://doi.org/10.1016/j.ijantimicag.2021.106421.9.
6) Morales, Jr, R., Kupa, L. V. K., Vianna, K. B., Garcia, M. S., Santos, V. J., Campos, E. V., Silva Jr, J. M., Silva Jr E. M., Oliveira, T. C., Gomez, D. S. & Santos, S. R. C. J. (2021). Meropenem extended infusion versus intermittent infusion against nosocomial MIC 4 mg/L strains to guarantee drug effectiveness by PK/PD approach in burn patients at the earlier period of septic shock. In *International Journal of Antimicrobial Agents*. Elsevier. https://doi.org/10.1016/j.ijantimicag.2021.106421.10.
7) Morales, Jr, R., Macedo, V. A. S., Pires, F. R., Campos, E. V., Ebner, P. A. R., Romano, P., Maria Severina dos Santos, Duarte, N. J., Gomez, D. S. & Santos, S. R. C. J. Vancomycin target attainment in critically ill septic pediatric burn versus non-burn patients against Staphylococcus spp. *International Journal of Antimicrobial Agents*, 58, 2100282. https://doi.org/10.1016/j.ijantimicag.2021.106421.11.
8) Morales, Jr, R., Chalom, M. Y., Romano, P., Ebner, P. A. R., Santos, M. S., Kupa, L. V. K., Sumita, N. M., Duarte N. J. C., Duarte, A. J. S., Gomez, D. S. & Santos, S. R. J. C. (2021). Meropenem 3 hrs-extended infusion improves effectiveness in critically ill burn patients at the earlier period of septic shock against P. aeruginosa intermediate susceptibility MIC 4-8

mg/L In *International Journal of Antimicrobial Agents.* Elsevier. https://doi.org/10.1016/j.ijantimicag.2021.106421.12.

9) Morales, Jr, R., Chalom, M. Y., Romano, P., Ebner, P. A. R., Santos, M. S., Kupa, L. V. K., Sumita, N. M., Duarte N. J. C., Duarte, A. J. S., Gomez, D. S. & Santos, S. R. J. C. (2021). Piperacillin/tazobactam effectiveness in septic burn patients requiring vasopressors by applying pharmacokinetics-pharmacodynamics approach. In *International Journal of Antimicrobial Agents.* Elsevier. https://doi.org/10.1016/j.ijantimicag.2021.106421.13.

10) Morales, Jr, R., Chalom, M. Y., Romano, P., Ebner, P. A. R., Santos, M. S., Garcia, M. S., Kupa, L. V. K., Duarte N. J. C., Duarte, A. J. S., Gomez, D. S. & Santos, S. R. J. C. (2021). Extended infusion improves meropenem and piperacillin/tazobactam effectiveness in septic burn patients with normal renal by applying pharmacokinetics-pharmacodynamics approach. In *International Journal of Antimicrobial Agents.* Elsevier. https://doi.org/10.1016/j.ijantimicag.2021.106421.14.

11) Messiano, C. G., Junior, R. M., Pereira, G. O., da Silva Junior, E. M., Gomez, D. S. & Santos, S. R. C. J. (2022). Therapeutic Target Attainment of 3-Hour Extended Infusion of Meropenem in Patients with Septic Burns. *Clinical Therapeutics*, 44(4), 624-629. http://dx.doi.org/10.1016/j.clinthera.2022.02.007.

12) de Camargo, T. V., Junior, E. M. S., Silva Jr, J. M., Gomez, D. S. & Santos, S. R. C. J. (2022). PK/PD approach to evaluate Meropenem effectiveness in critically ill burn adolescents versus young adults undergoing therapy of septic shock. *Pharm Pharmacol Int J*, 10(3), 79-85 (2022). https://doi.org/10.15406/ppij.2022.10.00368.

13) Santos, S. R. C. J., de Camargo T. V., Macedo V. A., Souza, V. K. & Gomez, D. S. Vancomycin and meropenem serum monitoring for target attainment by PK/PD approach as an effective tool in the battle against nosocomial pathogens in septic pediatric burn patients. *Pharm Pharmacol Int J.* 2022;10(3):87–95 (2022). https://doi.org/10.15406/ppij.2022.10.00369.

14) Santos, S. R. C. J., de Camargo T. V., Messiano C. G., Kupa L. V. K., Souza V. K., Ronaldo Morales Jr. R., Sanches Pinto D. C. S., Silva Junior E. M., Silva Junior J. M., Gomez D. S. Combating bacterial resistance to antimicrobials in severe septic ICU patients: importance of meropenem, piperacillin serum monitoring as a dose adjustment and duration of infusion strategies. *Pharm Pharmacol Int J.* 11(2):52–60 (2023). https://doi.org/10.15406/ppij.2023.11.00402.

15) Santos, S. R. C. J., de Camargo T. V., Messiano C. G., Souza V. K. Frederico Ribeiro Pires, Débora Cristina Sanches Pinto, Edvaldo Vieira de Campos, Elson Mendes da Silva Junior, João Manoel da Silva Junior, David de Souza Gomez. Combined vancomycin-meropenem therapy in pediatric major burns undergoing therapy of septic shock guided by cultures and pharmacokinetic-pharmacodynamics approach based on serum levels to combat bacterial resistance. *Pharm Pharmacol Int J.* 11(2):62–68 (2023). https://doi.org/10.15406/ppij.2023.11.00403.

Aline Sandré Gomides

ORCID https://orcid.org/0009-0004-8689-4208. Brazilian Platform of Research (CNPq).

Affiliation:

University of Sao Paulo – Nursing School (2008), Intensive Care Unit (2008-2023), Expertise in Intensive Care for Pediatric_Adult burn patients in Intensive Care Burns Unit.

Education:

University of São Paulo Nursing School_Expertise Nursing on turn in the Intensive Care Burn Unit of hospital Central Institute of Clinics Hospital of Medical School of University of Sao Paulo, since 2009.

Research Experience:

Head nurse on duty in ICU- Division of Plastic Surgery and Burns of Surgery Department of Medical School of University of São Paulo (2009-23).

Professional Experience:

Since 2008 actively participated in the implementation of the PK/PD approach based on serum levels of antimicrobials in the burn ICU for therapeutic drug monitoring related to antimicrobial effectiveness and safety. Individualization of therapy in ICU critically ill pediatric or adult burn patients started in 2007 at the laboratory of the Clinical Pharmacokinetics Center – University of Sao Paulo.

Professional Appointments:

Research fellow of team on Clinical Pharmacokinetics and Nurse on duty -in tertiary publica hospital of Medical School, University of Sao Paulo (2008-2023).

Honors: Scientific Awards from the Last 3 Years:

III International Meeting Sobratafe on Advanced Wound Care, SOBRATFE 2019, Maksoud Plaza Hotel São Paulo, SP. Awards At The Event Promoter: Brazilian Society for Advanced Treatment of Complex Wounds (Sobratafe) "Clinical Protocol"

Award 1: Vancomycin dose adjustment done in areal time for target attainment in critically ill septic pediatric burns versus non-burns against Staphylococcus spp MIC 1 mg/L. Camila Salinas Guillaux, Ronaldo Morales Júnior, Vedilaine Aparecida Bueno da Silva Macedo, Frederico Ribeiro Pires, Elson Mendes da Silva Júnior, João Manoel da Silva Júnior, Edvaldo Vieira de Campos, Amanda Maria Ribas Rosa de Oliveira, Aline Sandré Gomides, Gabriela Aparecida Ferreira, Thiago C. Oliveira, Silvia Regina Cavani Jorge Santos, David de Souza Gomez.

Award 2: Meropenem effectiveness in critically ill burn patients during the septic shock by pharmacokinetics assessment based on drug serum measurements done in a real time. Claudia Garcia Messiano, Elson Mendes da Silva Júnior, João Manoel da Silva Júnior, Edvaldo Vieira de Campos, Amanda Maria Ribas Rosa de Oliveira, Aline S. Gomides, Gabriela A. Ferreira, Thiago C. Oliveira, Silvia Regina Cavani Jorge Santos, David de Souza Gomez.

Brazilian Society of Infectiology (BSI) December/2020. The Brazilian Journal of Infectious Diseases, Volume 25, Supplement 1, January 2021. https://doi.org/10.1016/j.bjid.2020.101387. "Comparative study of the effectiveness of Piperacillin-Tazobactan after intermittent versus extended infusion in severe burn septic patients by the pharmacokinetic-pharmacodynamic approach (PK/PD)," PhD. student Vanessa Kasubeck de Souza, Aline Sandré Gomides, Gabriela Aparecida Ferreira, David de Souza Gomez & Silvia Regina Cavani Jorge Santos. https://doi.org/10.1016/j.bjid.2020.101387. ISSN 1413-86-70.

Research team member since 2008

Publications from the Last 3 Years:

1) Otofuji, G., Silva Jr, J. M., Silva Junior, E. M., Gomides, A. S., Rocha, A., Gomez, D. S. Lanchote V. L. & Santos, S. R. C. J. (2021). A alteração de conduta médica na prescrição de antimicrobiano guiada pelas culturas, biomarcadores e abordagem PK/PD de meropenem em paciente séptico grande queimado com infecção causada por p aeruginosa resistente à amicacina. *The Brazilian Journal of Infectious Diseases,* 25, 101409. https://doi.org/10.1016/j.bjid.2020.101409.
2) Matsuno, V. K., Junior, J. M. S., Gomides, A. S., Silva Filho, C. R., Santos, V. J., Rocha, A., Gomez, D. S. & Santos, S. R. C. J. (2021). abordagem PK/PD do fluconazol para garantir a efetividade da terapia antifúngica em pacientes sépticos grandes queimados em terapia intensiva com infecção sistêmica por Candida glabrata (CIM 32 mg/L). *The Brazilian Journal of Infectious Diseases,* 25, 101438. https://doi.org/10.1016/j.bjid.2020.101438.
3) Dos Santos, M. S., Silva Jr, J. M., Junior, E. M. S., Sandre A, Rocha A, Duarte, A. J. S., Gomez, D. S. Lanchote V. L. & Santos, S. R. C. J (2021). Efetividade da Terapia Antimicrobiana em paciente séptico grande queimado com disfunção renal e lesão ocular grave em terapia intensiva com vancomicina-piperacilina/tazobactana através da abordagem farmacocinética-farmacodinâmica. *The Brazilian Journal of Infectious Diseases*, 25, 101337. https://doi.org/10.1016/j.bjid.2020.101337.
4) Romano, P., Duarte, N. J. C., Silva Junior, J. M., Silva Junior, E. M., Sandre, A., Rocha, Duarte, A. J. S., A Gomez, D. S., Lanchote V. L. & Santos, S. R. C. J. (2021). Efetividade da terapia antimicrobiana em paciente séptico idoso grande queimado em terapia intensiva com vancomicina-meropenem através da abordagem farmacocinética-farmacodinâmica. *The Brazilian Journal of Infectious Diseases,* 25, 101403. https://doi.org/10.1016/j.bjid.2020.101403.
5) De Camargo T. C., Kupa, L. V. K., Silva Junior, J. M., Silva Junior, E. M, Oliveira, A. M. R. R., Oliveira, E. M., Gomides A. S., Ferreira G. A., Oliveira T. C., Gomez, D. S. & Santos, S. R. (2022). Alteração de conduta clínica em pacientes sépticos gravemente enfermos com queimaduras pelo ajuste da dose da vancomicina contra cepas gram-positivas CIM 2 mg/l pela abordagem farmacocinética-farmacodinâmica. *The Brazilian Journal of Infectious Diseases*, 26, 101738. https://doi.org/10.1016/j.bjid.2021.101738.

6) De Camargo, T. V., da Silva Jr, J. M. M., da Silva Junior, E. M., da Silva Filho, C. R., Santos, V. J., Oliveira, T. C., Gomez, D. S. & Santos, S. R. C. J. (2021). Estudo comparativo da efetividade do meropenem em pacientes sépticos queimados. adolescentes versus adultos jovens. *The Brazilian Journal of Infectious Diseases,* 25, 101388. https://doi.org/10.1016/j.bjid.2020.101388.
7) Morales Junior, R. M., Kupa, L. D., V. K., de Oliveira, E. M., de Campos, E. V., da Silva Junior, J. M., Junior, E. M. S., Gomez, D. S. & Santos, S. R. C. J. (2022). Cobertura antimicrobiana da vancomicina contra patógenos gram-positivos em pacientes queimados. *The Brazilian Journal of Infectious Diseases,* 26, 101745. https://doi.org/10.1016/j.bjid.2021.101745.
8) Messiano, C. G., Silva Junior, E. M., Silva Junior, J. M., Gomides, A. S., Ferreira, G. A., de Oliveira, T. C., David S. Gomez, & Santos, S. R. (2022). comparação da efetividade do meropenem nas fases precoce versus tardia do choque séptico através da abordagem farmacocinética-farmacodinâmica em pacientes queimados. *The Brazilian Journal of Infectious Diseases,* 26, 101746. https://doi.org/10.1016/j.bjid.2021.101746.
9) Morales, Jr., R., Kupa, L. V. K., Santos, V. J., Romano, P., Oliveira, E. M., Campos, E. V, Silva Jr, J. M., Silva Jr, E. M., Gomides A. S., Gomez, D. S. & Santos, S. R. C. J. (2021). Vancomycin dose adjustment against Gram-positive MIC 2 mg/L strains in critically ill adult burn patients by pharmacokinetic-pharmacodynamic approach. In *International Journal of Antimicrobial Agents.* Elsevier. https://doi.org/10.1016/j.ijantimicag.2021.106421.8.
10) Morales, Jr, R., Kupa, L. V. K., Silva Jr, J. M., Silva Jr, E. M., Oliveira, A. M. R. R., Gomides, A. S., Ferreira G. A., Gomez, D. S. & Santos, S. R. C. J. (2021). Meropenem extended infusion against nosocomial MIC 4 mg/L strains to guarantee drug effectiveness in burn patients at the earlier period of septic shock. *International Journal of Antimicrobial Agents,* 58, 21002806. https://doi.org/10.1016/j.ijantimicag.2021.106421.9.
11) Morales, Jr, R., Kupa, L. V. K., Vianna, K. B., Garcia, M. S., Santos, V. J., Campos, E. V., Silva Jr, J. M., Silva Jr E. M., Oliveira, T. C., Gomez, D. S. & Santos, S. R. C. J. (2021). Meropenem extended infusion versus intermittent infusion against nosocomial MIC 4 mg/L strains to guarantee drug effectiveness by PK/PD approach in burn patients at the earlier period of septic shock. In *International Journal of Antimicrobial Agents.* Elsevier. https://doi.org/10.1016/j.ijantimicag.2021.106421.10.

12) Morales, Jr, R., Macedo, V. A. S., Pires, F. R., Campos, E. V., Ebner, P. A. R., Romano, P., Maria Severina dos Santos, Duarte, N. J., Gomez, D. S. & Santos, S. R. C. J. Vancomycin target attainment in critically ill septic pediatric burn versus non-burn patients against Staphylococcus spp. *International Journal of Antimicrobial Agents*, 58, 2100282. https://doi.org/10.1016/j.ijantimicag.2021.106421.11.
13) Souza, V. K., Silva Jr E. M., Kaneko V. M., Silva Jr J. M., Gomides A. S., Ferreira G. A., Oliveira T. C., Gomez D. S., Santos, S. R. C. J. Extended two hours infusion improves piperacillin effectiveness against susceptible strains in septic burn patients. *Critical Care* (2019). v. 23(3): p. 21-22, https://doi.org/10.1186/s13054-019-9. ISSN 1364-8535.
14) Messiano, C. G., Junior, R. M., Pereira, G. O., da Silva Junior, E. M., Gomez, D. S. & Santos, S. R. C. J. (2022). Therapeutic Target Attainment of 3-Hour Extended Infusion of Meropenem in Patients with Septic Burns. *Clinical Therapeutics*, 44(4), 624-629. http://dx.doi.org/10.1016/j.clinthera.2022.02.007.
15) De Camargo, T. V., Junior, E. M. S., Silva Jr, J. M., Gomez, D. S. & Santos, S. R. C. J. (2022). PK/PD approach to evaluate Meropenem effectiveness in critically ill burn adolescents versus young adults undergoing therapy of septic shock. *Pharm Pharmacol Int J*, 10(3), 79-85 (2022). https://doi.org/10.15406/ppij.2022.10.00368.
16) Santos, S. R. C. J., de Camargo T. V., Macedo V. A., Souza, V. K. & Gomez, D. S. Vancomycin and meropenem serum monitoring for target attainment by PK/PD approach as an effective tool in the battle against nosocomial pathogens in septic pediatric burn patients. *Pharm Pharmacol Int J.* 2022;10(3):87–95 (2022). https://doi.org/10.15406/ppij.2022.10.00369.
17) Santos, S. R. C. J., de Camargo T. V., Messiano C. G., Kupa L. V. K., Souza V. K., Ronaldo Morales Jr. R., Sanches Pinto D. C. S., Silva Junior E. M., Silva Junior J. M., Gomez D. S. Combating bacterial resistance to antimicrobials in severe septic ICU patients: importance of meropenem, piperacillin serum monitoring as a dose adjustment and duration of infusion strategies. *Pharm Pharmacol Int J.* 11(2):52–60 (2023). https://doi.org/10.15406/ppij.2023.11.00402.
18) Santos, S. R. C. J., de Camargo T. V., Messiano C. G., Souza V. K. Frederico Ribeiro Pires, Débora Cristina Sanches Pinto, Edvaldo Vieira de Campos, Elson Mendes da Silva Junior, João Manoel da Silva Junior, David de Souza Gomez. Combined vancomycin-meropenem therapy in pediatric major burns undergoing therapy of septic shock guided by

cultures and pharmacokinetic-pharmacodynamics approach based on serum levels to combat bacterial resistance. *Pharm Pharmacol Int J.* 11(2):62–68 (2023). https://doi.org/10.15406/ppij.2023.11.00403.

Gabriela Aparecida Ferreira

ORCID https://orcid.org/0009-0008-3798-0504. Brazilian Platform of Research (CNPq).

Affiliation:
University of Sao Paulo – Nursing School (2008), Intensive Care Unit (2008-2023), Expertise in Intensive Care for Pediatric_Adult burn patients in Intensive Care Burns Unit.

Education:
University of São Paulo Nursing School_Expertise Nursing on turn in the Intensive Care Burn Unit of hospital Central Institute of Clinics Hospital of Medical School of University of Sao Paulo, since 2008.

Research Experience:
Head nurse on duty in ICU- Division of Plastic Surgery and Burns of Surgery Department of Medical School of University of São Paulo (2009-23).

Professional Experience:
Since 2008 actively participated in the implementation of the PK/PD approach based on serum levels of antimicrobials in the burn ICU for therapeutic drug monitoring related to antimicrobial effectiveness and safety. Individualization of therapy in ICU critically ill pediatric or adult burn patients started in 2007 at the laboratory of the Clinical Pharmacokinetics Center – University of Sao Paulo.

Professional Appointments:
Research fellow of team on Clinical Pharmacokinetics and Nurse on duty -in tertiary publica hospital of Medical School, University of Sao Paulo (2008-2023).

Honors: Scientific Awards from the Last 3 Years:

III International Meeting Sobratafe on Advanced Wound Care, Sobratfe 2019, Maksoud Plaza Hotel São Paulo, SP. Awards at The Event Promoter: Brazilian Society for Advanced Treatment of Complex Wounds (Sobratafe) "Clinical Protocol"

Award 1: Vancomycin dose adjustment done in areal time for target attainment in critically ill septic pediatric burns versus non-burns against Staphylococcus spp MIC 1 mg/L. Camila Salinas Guillaux, Ronaldo Morales Júnior, Vedilaine Aparecida Bueno da Silva Macedo, Frederico Ribeiro Pires, Elson Mendes da Silva Júnior, João Manoel da Silva Júnior, Edvaldo Vieira de Campos, Amanda Maria Ribas Rosa de Oliveira, Aline Sandré Gomides, Gabriela Aparecida Ferreira, Thiago C. Oliveira, Silvia Regina Cavani Jorge Santos, David de Souza Gomez.

Award 2: Meropenem effectiveness in critically ill burn patients during the septic shock by pharmacokinetics assessment based on drug serum measurements done in a real time. Claudia Garcia Messiano, Elson Mendes da Silva Júnior, João Manoel da Silva Júnior, Edvaldo Vieira de Campos, Amanda Maria Ribas Rosa de Oliveira, Aline S. Gomides, Gabriela A. Ferreira, Thiago C. Oliveira, Silvia Regina Cavani Jorge Santos, David de Souza Gomez.

Brazilian Society of Infectiology (BSI) December/2020. *The Brazilian Journal of Infectious Diseases,* Volume 25, Supplement 1, January 2021 ISSN 1413-86-70. https://doi.org/10.1016/j.bjid.2020.101387. "Comparative study of the effectiveness of Piperacillin-Tazobactam after intermittent versus extended infusion in severe burn septic patients by the pharmacokinetic-pharmacodynamic approach (PK/PD)," Master of Sciences. student Vanessa Kasubeck de Souza, Aline Sandré Gomides, Gabriela Aparecida Ferreira, David de Souza Gomez & Silvia Regina Cavani Jorge Santos.

Research team member since 2008

Publications from the Last 3 Years:

1) Souza, V. K., Silva Jr, J. M., Junior, E. M. S., Ferreira, G. A., Silva Filho, C. R., Santos, V. J., Gomez, D. S. & Santos, S. R. C. J (2021). Estudo comparativo da efetividade da piperacilina-tazobactana após infusão intermitente versus estendida em pacientes sépticos grandes queimados pela abordagem farmacocinética-farmacodinâmica (PK/PD). *The*

Brazilian Journal of Infectious Diseases, 25, 101387. https://doi.org/10.1016/j.bjid.2020.101387.

2) de Camargo, T. V., da Silva Jr, J. M. M., da Silva Junior, E. M., da Silva Filho, C. R., Santos, V. J., Oliveira, T. C., Gomez, D. S. & Santos, S. R. C. J. (2021). Estudo comparativo da efetividade do meropenem em pacientes sépticos queimados. adolescentes versus adultos jovens. *The Brazilian Journal of Infectious Diseases,* 25, 101388. https://doi.org/10.1016/j.bjid.2020.101388.
3) Souza V. K., Silva Jr J. M., Silva Jr E. M., Ferreira, G. A., Silva Filho, C. R., Santos, V. J., Rocha A., Lanchote, V. L., Gomez D. S., Santos S. R. C. J. Estudo comparativo da efetividade da piperacilina-tazobactana após infusão intermitente versus estendida em pacientes sépticos grandes queimados pela abordagem farmacocinética-farmacodinâmica (PK/PD) *The Brazilian Journal of Infectious Diseases,* 26,101337. https://doi.org/10.1016/j.bjid.2021.101337.
4) Otofuji, G., Morales Junior, R, Silva Junior, J. M., Silva Junior E. M., Oliveira, A. M. R. R., Oliveira, E. M., Ferreira G. A., Oliveira T. C., Gomez, D. S & Santos, S. R. (2022). Abordagem PK/PD na efetividade de piperacilina-vancomicina em pacientes sépticos queimados em regime empírico da dose recomendada na insuficiência renal. *The Brazilian Journal of Infectious Diseases,* 26, 101737. https://doi.org/10.1016/j.bjid.2021.101737.
5) de Camargo T. C., Kupa, L. V. K., Silva Junior, J. M., Silva Junior, E. M, Oliveira, A. M. R. R., Oliveira, E. M., Gomides A. S., Ferreira G. A., Oliveira T. C., Gomez, D. S. & Santos, S. R. (2022). Alteração de conduta clínica em pacientes sépticos gravemente enfermos com queimaduras pelo ajuste da dose da vancomicina contra cepas gram-positivas CIM 2 mg/l pela abordagem farmacocinética-farmacodinâmica. *The Brazilian Journal of Infectious Diseases*, 26, 101738. https://doi.org/10.1016/j.bjid.2021.101738.
6) de Camargo, T. V., da Silva Jr, J. M. M., da Silva Junior, E. M., da Silva Filho, C. R., Santos, V. J., Oliveira, T. C., Gomez, D. S. & Santos, S. R. C. J. (2021). Estudo comparativo da efetividade do meropenem em pacientes sépticos queimados. adolescentes versus adultos jovens. *The Brazilian Journal of Infectious Diseases,* 25, 101388. https://doi.org/10.1016/j.bjid.2020.101388.
7) Morales Junior, R. M., Kupa, L. D., V. K., de Oliveira, E. M., de Campos, E. V., da Silva Junior, J. M., Junior, E. M. S., Gomez, D. S. & Santos, S. R. C. J. (2022). Cobertura antimicrobiana da vancomicina contra

patógenos gram-positivos em pacientes queimados. *The Brazilian Journal of Infectious Diseases,* 26, 101745. https://doi.org/10.1016/j.bjid.2021.101745.

8) Messiano, C. G., Silva Junior, E. M., Silva Junior, J. M., Gomides, A. S., Ferreira, G. A., de Oliveira, T. C., David S. Gomez, & Santos, S. R. (2022). comparação da efetividade do meropenem nas fases precoce versus tardia do choque séptico através da abordagem farmacocinética-farmacodinâmica em pacientes queimados. *The Brazilian Journal of Infectious Diseases,* 26, 101746. https://doi.org/10.1016/j.bjid.2021.101746.
9) Morales, Jr, R., Kupa, L. V. K., Silva Jr, J. M., Silva Jr, E. M., Oliveira, A. M. R. R., Gomides, A. S., Ferreira G. A., Gomez, D. S. & Santos, S. R. C. J. (2021). Meropenem extended infusion against nosocomial MIC 4 mg/L strains to guarantee drug effectiveness in burn patients at the earlier period of septic shock. *International Journal of Antimicrobial Agents,* 58, 21002806. https://doi.org/10.1016/j.ijantimicag.2021.106421.9.
10) Morales, Jr, R., Kupa, L. V. K., Vianna, K. B., Garcia, M. S., Santos, V. J., Campos, E. V., Silva Jr, J. M., Silva Jr E. M., Oliveira, T. C., Gomez, D. S. & Santos, S. R. C. J. (2021). Meropenem extended infusion versus intermittent infusion against nosocomial MIC 4 mg/L strains to guarantee drug effectiveness by PK/PD approach in burn patients at the earlier period of septic shock. In *International Journal of Antimicrobial Agents*. Elsevier. https://doi.org/10.1016/j.ijantimicag.2021.106421.10.
11) Morales, Jr, R., Macedo, V. A. S., Pires, F. R., Campos, E. V., Ebner, P. A. R., Romano, P., Maria Severina dos Santos, Duarte, N. J., Gomez, D. S. & Santos, S. R. C. J. Vancomycin target attainment in critically ill septic pediatric burn versus non-burn patients against Staphylococcus spp. *International Journal of Antimicrobial Agents*, 58, 2100282. https://doi.org/10.1016/j.ijantimicag.2021.106421.11.
12) Souza, V. K., Silva Jr E. M., Kaneko V. M., Silva Jr J. M., Gomides A. S., Ferreira G. A., Oliveira T. C., Gomez D. S., Santos, S. R. C. J. Extended two hours infusion improves piperacillin effectiveness against susceptible strains in septic burn patients. *Critical Care* (2019). v. 23(3): p 21-22, https://doi.org/10.1186/s13054-019-9. ISSN 1364-8535.
13) Messiano, C. G., Junior, R. M., Pereira, G. O., da Silva Junior, E. M., Gomez, D. S. & Santos, S. R. C. J. (2022). Therapeutic Target Attainment of 3-Hour Extended Infusion of Meropenem in Patients with Septic Burns. *Clinical Therapeutics*, 44(4), 624-629. http://dx.doi.org/10.1016/j.clinthera.2022.02.007.

14) de Camargo, T. V., Junior, E. M. S., Silva Jr, J. M., Gomez, D. S. & Santos, S. R. C. J. (2022). PK/PD approach to evaluate Meropenem effectiveness in critically ill burn adolescents versus young adults undergoing therapy of septic shock. *Pharm Pharmacol Int J*, 10(3), 79-85 (2022). https://doi.org/10.15406/ppij.2022.10.00368.
15) Santos, S. R. C. J., de Camargo T. V., Macedo V. A., Souza, V. K. & Gomez, D. S. Vancomycin and meropenem serum monitoring for target attainment by PK/PD approach as an effective tool in the battle against nosocomial pathogens in septic pediatric burn patients. *Pharm Pharmacol Int J.* 2022;10(3):87–95 (2022). https://doi.org/10.15406/ppij.2022.10.00369.
16) Santos, S. R. C. J., de Camargo T. V., Messiano C. G., Kupa L. V. K., Souza V. K., Ronaldo Morales Jr. R., Sanches Pinto D. C. S., Silva Junior E. M., Silva Junior J. M., Gomez D. S. Combating bacterial resistance to antimicrobials in severe septic ICU patients: importance of meropenem, piperacillin serum monitoring as a dose adjustment and duration of infusion strategies. *Pharm Pharmacol Int J.* 11(2):52–60 (2023). https://doi.org/10.15406/ppij.2023.11.00402.
17) Santos, S. R. C. J., de Camargo T. V., Messiano C. G., Souza V. K. Frederico Ribeiro Pires, Débora Cristina Sanches Pinto, Edvaldo Vieira de Campos, Elson Mendes da Silva Junior, João Manoel da Silva Junior, David de Souza Gomez. Combined vancomycin-meropenem therapy in pediatric major burns undergoing therapy of septic shock guided by cultures and pharmacokinetic-pharmacodynamics approach based on serum levels to combat bacterial resistance. *Pharm Pharmacol Int J.* 11(2):62–68 (2023). https://doi.org/10.15406/ppij.2023.11.00403.

Edvaldo Vieira de Campos

ORCID https://orcid.org/0000-0002-7343-4526. Brazilian Platform of Research (CNPq).

Affiliation:

University of Sao Paulo – Medical Doctor of Emergency Care (2010), Expertise Intensive Care Unit (2007-2023), Expertise in Intensive Care for Adult burn patients in Intensive Care Burns Unit (2007-2016).

Education:

University Federal of Maringá, PR. Expertise in the Intensive Care in Burns, Intensive Care Unit of hospital, Central Institute of Clinics Hospital, Medical School of University of Sao Paulo, since 2006. Medical Doctor (2010).

Research Experience:

Intensive on duty in ICU- Division of Plastic Surgery and Burns of Surgery Department of Medical School of University of São Paulo (2006-2023).

Professional Experience:

Since 2007 actively participated in the implementation of the PK/PD approach based on serum levels of antimicrobials in the burn ICU for therapeutic drug monitoring related to antimicrobial effectiveness and safety. Individualization of therapy in ICU critically ill pediatric and adult burn patients started in 2007 at the laboratory of the Clinical Pharmacokinetics Center – University of Sao Paulo.

Professional Appointments:

Research fellow of team on Clinical Pharmacokinetics and Intensive on duty -in tertiary public hospital of Medical School, University of Sao Paulo (2007-2023).

Honors: Scientific Awards from the Last 3 Years:

III International Meeting Sobratafe on Advanced Wound Care, Sobratfe 2019, Maksoud Plaza Hotel São Paulo, SP. Awards At The Event Promoter: Brazilian Society for Advanced Treatment of Complex Wounds (SOBRATAFE) "Clinical Protocol"

Award 1: Vancomycin dose adjustment done in areal time for target attainment in critically ill septic pediatric burns versus non-burns against Staphylococcus spp MIC 1 mg/L. Camila Salinas Guillaux, Ronaldo Morales Júnior, Vedilaine Aparecida Bueno da Silva Macedo, Frederico Ribeiro Pires, Elson Mendes da Silva Júnior, João Manoel da Silva Júnior, Edvaldo Vieira de Campos, Amanda Maria Ribas Rosa de Oliveira, Aline Sandré Gomides, Gabriela Aparecida Ferreira, Thiago C. Oliveira, Silvia Regina Cavani Jorge Santos, David de Souza Gomez.

Award 2: Meropenem effectiveness in critically ill burn patients during the septic shock by pharmacokinetics assessment based on drug serum measurements done in a real time. Claudia Garcia Messiano, Elson Mendes da Silva Júnior, João Manoel da Silva Júnior, Edvaldo Vieira de Campos, Amanda Maria Ribas Rosa de Oliveira, Aline S. Gomides, Gabriela A. Ferreira, Thiago C. Oliveira, Silvia Regina Cavani Jorge Santos, David de Souza Gomez.

National Meeting Sobratafe on Advanced Wound Care, Sobratfe 2020, Maksoud Plaza Hotel São Paulo, SP. Four Awards at The Event. Promoter: Brazilian Society for Advanced Treatment of Complex Wounds Dr Walter Soares Pinto Prize

Award 1: Is Meropenem effectiveness in ICU burn patients during SIRS at the earlier period of septic shock impacted by pharmacokinetic changes? Adolescents versus Young adults. Thais V. de Camargo, Edvaldo Vieira de Campos, Amanda Maria Ribas Rosa de Oliveira, João Manoel da Silva Júnior, Elson Mendes da Silva Júnior, Aline S. Gomides, Gabriela A. Ferreira, Thiago C. Oliveira, Vera Lúcia Lanchote, Silvia Regina Cavani Jorge Santos, David de Souza Gomez.

Award 2: Extended infusion improves beta-lactams effectiveness in septic burn patients with normal renal function against P aeruginosa intermediate susceptibility. Vanessa Kasubeck de Souza, Karina B Vianna, João Manoel da Silva Júnior, Amanda Maria Ribas Rosa de Oliveira, Edvaldo Vieira de Campos, Elson Mendes da Silva Júnior, Aline S. Gomides, Gabriela A. Ferreira, Thiago C. Oliveira, Vera Lúcia Lanchote, Silvia Regina Cavani Jorge Santos, David de Souza Gomez.

Award 3: Pharmacokinetic-pharmacodynamic approach permits vancomycin dose adjustment for effectiveness against infection caused by Gram-positive pathogens MIC > 1 mg/L in ICU paediatric burn patients. Ronaldo Morales Júnior, Thais Vieira de Camargo, Vedilaine Aparecida Bueno da Silva Macedo, Edvaldo Vieira de Campos, João Manoel da Silva Júnior, Nilo José Coelho Duarte, Silvia Regina Cavani Jorge Santos, David de Souza Gomez.

Apfid-Prize (Asian Pacific Foundation For Infectious Diseases) "2021 Outstanding Poster Award" Leonard de Vinci Kanda Kupa, Ronaldo *Morales Jr*, Aline Sandré Gomides, Gabriela Aparecida Ferreira, Edvaldo Vieira de Campos, João Manoel da Silva Júnior, Elson Mendes da Silva Junior, David de Souza Gomez & Silvia R. C. J. Santos. https://doi.org/10.1016/j.ijantimicag.2021.106421.10.

Research team member since 2007

Publications from the Last 3 Years:

1) Souza, V. K., Silva Jr, J. M., Junior, E. M. S., Ferreira, G. A., Silva Filho, C. R., Santos, V. J., Gomez, D. S. & Santos, S. R. C. J (2021). Estudo comparativo da efetividade da piperacilina-tazobactana após infusão intermitente versus estendida em pacientes sépticos grandes queimados pela abordagem farmacocinética-farmacodinâmica (PK/PD). *The Brazilian Journal of Infectious Diseases*, 25, 101387. https://doi.org/10.1016/j.bjid.2020.101387.
2) Souza V. K., Silva Jr J. M., Silva Jr E. M., Ferreira, G. A., Silva Filho, C. R., Santos, V. J., Rocha A., Lanchote, V. L., Gomez D. S., Santos S. R. C. J. Estudo comparativo da efetividade da piperacilina-tazobactana após infusão intermitente versus estendida em pacientes sépticos grandes queimados pela abordagem farmacocinética-farmacodinâmica (PK/PD) *The Brazilian Journal of Infectious Diseases,* 26,101337. https://doi.org/10.1016/j.bjid.2021.101337.
3) Morales Junior, R. M., Kupa, L. D., V. K., de Oliveira, E. M., de Campos, E. V., da Silva Junior, J. M., Junior, E. M. S., Gomez, D. S. & Santos, S. R. C. J. (2022). Cobertura antimicrobiana da vancomicina contra patógenos gram-positivos em pacientes queimados. *The Brazilian Journal of Infectious Diseases,* 26, 101745. https://doi.org/10.1016/j.bjid.2021.101745.
4) Morales, Jr., R., Kupa, L. V. K., Santos, V. J., Romano, P., Oliveira, E. M., Campos, E. V, Silva Jr, J. M., Silva Jr, E. M., Gomides A. S., Gomez, D. S. & Santos, S. R. C. J. (2021). Vancomycin dose adjustment against Gram-positive MIC 2 mg/L strains in critically ill adult burn patients by pharmacokinetic-pharmacodynamic approach. In *International Journal of Antimicrobial Agents.* Elsevier. https://doi.org/10.1016/j.ijantimicag.2021.106421.8.
5) Morales, Jr, R., Kupa, L. V. K., Silva Jr, J. M., Silva Jr, E. M., Oliveira, A. M. R. R., Gomides, A. S., Ferreira G. A., Gomez, D. S. & Santos, S. R. C. J. (2021). Meropenem extended infusion against nosocomial MIC 4 mg/L strains to guarantee drug effectiveness in burn patients at the earlier period of septic shock. *International Journal of Antimicrobial Agents,* 58, 21002806. https://doi.org/10.1016/j.ijantimicag.2021.106421.9.
6) Morales, Jr, R., Kupa, L. V. K., Vianna, K. B., Garcia, M. S., Santos, V. J., Campos, E. V., Silva Jr, J. M., Silva Jr E. M., Oliveira, T. C., Gomez,

D. S. & Santos, S. R. C. J. (2021). Meropenem extended infusion versus intermittent infusion against nosocomial MIC 4 mg/L strains to guarantee drug effectiveness by PK/PD approach in burn patients at the earlier period of septic shock. In *International Journal of Antimicrobial Agents*. Elsevier. https://doi.org/10.1016/j.ijantimicag.2021.106421.10.

7) Morales, Jr, R., Macedo, V. A. S., Pires, F. R., Campos, E. V., Ebner, P. A. R., Romano, P., Maria Severina dos Santos, Duarte, N. J., Gomez, D. S. & Santos, S. R. C. J. Vancomycin target attainment in critically ill septic pediatric burn versus non-burn patients against Staphylococcus spp. *International Journal of Antimicrobial Agents*, 58, 2100282. https://doi.org/10.1016/j.ijantimicag.2021.106421.11.
8) Souza, V. K., Silva Jr E. M., Kaneko V. M., Silva Jr J. M., Gomides A. S., Ferreira G. A., Oliveira T. C., Gomez D. S., Santos, S. R. C. J. Extended two hours infusion improves piperacillin effectiveness against susceptible strains in septic burn patients. *Critical Care* (2019). v. 23(3): p 21-22, https://doi.org/10.1186/s13054-019-9. ISSN 1364-8535.
9) Messiano, C. G., Junior, R. M., Pereira, G. O., da Silva Junior, E. M., Gomez, D. S. & Santos, S. R. C. J. (2022). Therapeutic Target Attainment of 3-Hour Extended Infusion of Meropenem in Patients with Septic Burns. *Clinical Therapeutics*, 44(4), 624-629. http://dx.doi.org/10.1016/j.clinthera.2022.02.007.
10) De Camargo, T. V., Junior, E. M. S., Silva Jr, J. M., Gomez, D. S. & Santos, S. R. C. J. (2022). PK/PD approach to evaluate Meropenem effectiveness in critically ill burn adolescents versus young adults undergoing therapy of septic shock. *Pharm Pharmacol Int J*, 10(3), 79-85 (2022). https://doi.org/10.15406/ppij.2022.10.00368.
11) Santos, S. R. C. J., de Camargo T. V., Macedo V. A., Souza, V. K. & Gomez, D. S. Vancomycin and meropenem serum monitoring for target attainment by PK/PD approach as an effective tool in the battle against nosocomial pathogens in septic pediatric burn patients. *Pharm Pharmacol Int J.* 2022;10(3):87–95 (2022). https://doi.org/10.15406/ppij.2022.10.00369.
12) Santos, S. R. C. J., de Camargo T. V., Messiano C. G., Souza V. K. Frederico Ribeiro Pires, Débora Cristina Sanches Pinto, Edvaldo Vieira de Campos, Elson Mendes da Silva Junior, João Manoel da Silva Junior, David de Souza Gomez. Combined vancomycin-meropenem therapy in pediatric major burns undergoing therapy of septic shock guided by cultures and pharmacokinetic-pharmacodynamics approach based on

serum levels to combat bacterial resistance. *Pharm Pharmacol Int J.* 11(2):62–68 (2023). https://doi.org/10.15406/ppij.2023.11.00403.

João Manoel da Silva Junior,
Associate Professor

ORCID https://orcid.org/0000-0003-1494-0784. Brazilian Platform of Research (CNPq).

Affiliation:
University of Sao Paulo – Medical Doctor of Anesthesiology (2009), Expertise Intensive Care Unit (2007-2023), Expertise in Intensive Care for adult burn patients in Intensive Care Burns Unit (2007-2022).

Education:
University of Sao Paulo, SP. Expertise in the Intensive Care in Burns - Intensive Care Unit of hospital Central Institute of Clinics Hospital of Medical School of University of Sao Paulo, since 2009. Medical Doctor (2009).

Research Experience:
Intensive on duty in ICU- Division of Plastic Surgery and Burns of Surgery Department of Medical School of University of São Paulo (2007-2022).

Professional Experience:
Since 2009 actively participated in the implementation of the PK/PD approach based on serum levels of antimicrobials in the burn ICU for therapeutic drug monitoring related to antimicrobial effectiveness and safety. Individualization of therapy in ICU critically ill pediatric and adult burn patients started in 2007 at the laboratory of the Clinical Pharmacokinetics Center – University of Sao Paulo.

Professional Appointments:
Research fellow of team on Clinical Pharmacokinetics in tertiary public hospital of Medical School, University of Sao Paulo (2009-2023). Lecture of Medical School, University of Sao Paulo (2021). Associate Professor of Medical School of University of Sao Paulo (2022).

Honors: Scientific Awards from the Last 3 Years:

III International Meeting Sobratafe on Advanced Wound Care, Sobratfe 2019, Maksoud Plaza Hotel São Paulo, SP. Awards at the Event Promoter: Brazilian Society for Advanced Treatment of Complex Wounds (Sobratafe)

"Clinical Protocol" Award 1: Vancomycin dose adjustment done in areal time for target attainment in critically ill septic pediatric burns versus non-burns against Staphylococcus spp MIC 1 mg/L. Camila Salinas Guillaux, Ronaldo Morales Júnior, Vedilaine Aparecida Bueno da Silva Macedo, Frederico Ribeiro Pires, Elson Mendes da Silva Júnior, João Manoel da Silva Júnior, Edvaldo Vieira de Campos, Amanda Maria Ribas Rosa de Oliveira, Aline Sandré Gomides, Gabriela Aparecida Ferreira, Thiago C. Oliveira, Silvia Regina Cavani Jorge Santos, David de Souza Gomez.

"Clinical Protocol" Award 2: Meropenem effectiveness in critically ill burn patients during the septic shock by pharmacokinetics assessment based on drug serum measurements done in a real time. Claudia Garcia Messiano, Elson Mendes da Silva Júnior, João Manoel da Silva Júnior, Edvaldo Vieira de Campos, Amanda Maria Ribas Rosa de Oliveira, Aline S. Gomides, Gabriela A. Ferreira, Thiago C. Oliveira, Silvia Regina Cavani Jorge Santos, David de Souza Gomez.

National Meeting Sobratafe on Advanced Wound Care, Sobratfe 2020, Maksoud Plaza Hotel São Paulo, SP. Four Awards at The Event. Promoter: Brazilian Society for Advanced Treatment of Complex Wounds "Dr Walter Soares Pinto Prize"

Award 1: Is Meropenem effectiveness in ICU burn patients during SIRS at the earlier period of septic shock impacted by pharmacokinetic changes? Adolescents versus Young adults. Thais V. de Camargo, Edvaldo Vieira de Campos, Amanda Maria Ribas Rosa de Oliveira, João Manoel da Silva Júnior, Elson Mendes da Silva Júnior, Aline S. Gomides, Gabriela A. Ferreira, Thiago C. Oliveira, Vera Lúcia Lanchote, Silvia Regina Cavani Jorge Santos, David de Souza Gomez.

Award 2: Extended infusion improves beta-lactams effectiveness in septic burn patients with normal renal function against P aeruginosa intermediate susceptibility. Vanessa Kasubeck de Souza, Karina B Vianna, João Manoel da Silva Júnior, Amanda Maria Ribas Rosa de Oliveira, Edvaldo Vieira de Campos, Elson Mendes da Silva Júnior, Aline S. Gomides, Gabriela A. Ferreira, Thiago C. Oliveira, Vera Lúcia Lanchote, Silvia Regina Cavani Jorge Santos, David de Souza Gomez.

Award 3: Pharmacokinetic-pharmacodynamic approach permits vancomycin dose adjustment for effectiveness against infection caused by Gram-positive pathogens MIC > 1 mg/L in ICU paediatric burn patients. Ronaldo Morales Júnior, Thais Vieira de Camargo, Vedilaine Aparecida Bueno da Silva Macedo, Edvaldo Vieira de Campos, João Manoel da Silva Júnior, Nilo José Coelho Duarte, Silvia Regina Cavani Jorge Santos, David de Souza Gomez.

Apfid-Prize (Asian Pacific Foundation For Infectious Diseases) "2021 Outstanding Poster Award" Leonard de Vinci Kanda Kupa, R. *Morales, Jr*, Edvaldo Vieira de Campos, João Manoel da Silva Júnior, Elson Mendes da Silva Junior, David de Souza Gomez & Silvia R. C. J. Santos. https://doi.org/10.1016/j.ijantimicag.2021.106421.10.

Research team member since 2009-2023

Publications from the Last 3 Years

1) Souza, V. K., Silva Jr, J. M., Junior, E. M. S., Ferreira, G. A., Silva Filho, C. R., Santos, V. J., Gomez, D. S. & Santos, S. R. C. J (2021). Estudo comparativo da efetividade da piperacilina-tazobactana após infusão intermitente versus estendida em pacientes sépticos grandes queimados pela abordagem farmacocinética-farmacodinâmica (PK/PD). *The Brazilian Journal of Infectious Diseases*, 25, 101387. https://doi.org/10.1016/j.bjid.2020.101387.
2) Souza V. K., Silva Jr J. M., Silva Jr E. M., Ferreira, G. A., Silva Filho, C. R., Santos, V. J., Rocha A., Lanchote, V. L., Gomez D. S., Santos S. R. C. J. Estudo comparativo da efetividade da piperacilina-tazobactana após infusão intermitente versus estendida em pacientes sépticos grandes queimados pela abordagem farmacocinética-farmacodinâmica (PK/PD) *The Brazilian Journal of Infectious Diseases,* 26,101337. https://doi.org/10.1016/j.bjid.2021.101337.
3) Morales Junior, R. M., Kupa, L. D., V. K., de Oliveira, E. M., de Campos, E. V., da Silva Junior, J. M., Junior, E. M. S., Gomez, D. S. & Santos, S. R. C. J. (2022). Cobertura antimicrobiana da vancomicina contra patógenos gram-positivos em pacientes queimados. *The Brazilian Journal of Infectious Diseases,* 26, 101745. https://doi.org/10.1016/j.bjid.2021.101745.
4) Morales, Jr., R., Kupa, L. V. K., Santos, V. J., Romano, P., Oliveira, E. M., Campos, E. V, Silva Jr, J. M., Silva Jr, E. M., Gomides A. S., Gomez,

D. S. & Santos, S. R. C. J. (2021). Vancomycin dose adjustment against Gram-positive MIC 2 mg/L strains in critically ill adult burn patients by pharmacokinetic-pharmacodynamic approach. In *International Journal of Antimicrobial Agents*. Elsevier. https://doi.org/10.1016/j.ijantimicag.2021.106421.8.

5) Morales, Jr, R., Kupa, L. V. K., Silva Jr, J. M., Silva Jr, E. M., Oliveira, A. M. R. R., Gomides, A. S., Ferreira G. A., Gomez, D. S. & Santos, S. R. C. J. (2021). Meropenem extended infusion against nosocomial MIC 4 mg/L strains to guarantee drug effectiveness in burn patients at the earlier period of septic shock. *International Journal of Antimicrobial Agents,* 58, 21002806. https://doi.org/10.1016/j.ijantimicag.2021.106421.9.
6) Morales, Jr, R., Kupa, L. V. K., Vianna, K. B., Garcia, M. S., Santos, V. J., Campos, E. V., Silva Jr, J. M., Silva Jr E. M., Oliveira, T. C., Gomez, D. S. & Santos, S. R. C. J. (2021). Meropenem extended infusion versus intermittent infusion against nosocomial MIC 4 mg/L strains to guarantee drug effectiveness by PK/PD approach in burn patients at the earlier period of septic shock. In *International Journal of Antimicrobial Agents*. Elsevier. https://doi.org/10.1016/j.ijantimicag.2021.106421.10.
7) Souza, V. K., Silva Jr E. M., Kaneko V. M., Silva Jr J. M., Gomides A. S., Ferreira G. A., Oliveira T. C., Gomez D. S., Santos, S. R. C. J. Extended two hours infusion improves piperacillin effectiveness against susceptible strains in septic burn patients. *Critical Care* (2019). v. 23(3): p 21-22, https://doi.org/10.1186/s13054-019-9. ISSN 1364-8535.
8) Messiano, C. G., Junior, R. M., Pereira, G. O., da Silva Junior, E. M., Gomez, D. S. & Santos, S. R. C. J. (2022). Therapeutic Target Attainment of 3-Hour Extended Infusion of Meropenem in Patients with Septic Burns. *Clinical Therapeutics*, 44(4), 624-629. http://dx.doi.org/10.1016/j.clinthera.2022.02.007.
9) de Camargo, T. V., Junior, E. M. S., Silva Jr, J. M., Gomez, D. S. & Santos, S. R. C. J. (2022). PK/PD approach to evaluate Meropenem effectiveness in critically ill burn adolescents versus young adults undergoing therapy of septic shock. *Pharm Pharmacol Int J*, 10(3), 79-85 (2022). https://doi.org/10.15406/ppij.2022.10.00368.
10) Santos, S. R. C. J., de Camargo T. V., Macedo V. A., Souza, V. K. & Gomez, D. S. Vancomycin and meropenem serum monitoring for target attainment by PK/PD approach as an effective tool in the battle against nosocomial pathogens in septic pediatric burn patients. *Pharm Pharmacol Int J.* 2022;10(3):87–95 (2022). https://doi.org/10.15406/ppij.2022.10.00369.

11) Santos, S. R. C. J., de Camargo T. V., Messiano C. G., Kupa L. V. K., Souza V. K., Ronaldo Morales Jr. R., Sanches Pinto D. C. S., Silva Junior E. M., Silva Junior J. M., Gomez D. S. Combating bacterial resistance to antimicrobials in severe septic ICU patients: importance of meropenem, piperacillin serum monitoring as a dose adjustment and duration of infusion strategies. *Pharm Pharmacol Int J.* 11(2):52–60 (2023). https://doi.org/10.15406/ppij.2023.11.00402.
12) Santos, S. R. C. J., de Camargo T. V., Messiano C. G., Souza V. K. Frederico Ribeiro Pires, Débora Cristina Sanches Pinto, Edvaldo Vieira de Campos, Elson Mendes da Silva Junior, João Manoel da Silva Junior, David de Souza Gomez. Combined vancomycin-meropenem therapy in pediatric major burns undergoing therapy of septic shock guided by cultures and pharmacokinetic-pharmacodynamics approach based on serum levels to combat bacterial resistance. *Pharm Pharmacol Int J.* 11(2):62–68 (2023). https://doi.org/10.15406/ppij.2023.11.00403.

Elson Mendes da Silva Junior

ORCID https://orcid.org/0000-0001-9112-8306. Brazilian Platform of Research (CNPq).

Affiliation:

University of Sao Paulo, doctorand student since 2021 of the Graduate Program "Anesthesiology, Surgical Sciences and Perioperative Medicine," Protocol CAPPESQ N. CAAE 07525118.3.0000.0068. Intensive Care Unit of Burns: Expertise Intensive Care Unit of Central Institute of Clinics Hospital of Medical School, University of Sao Paulo/SP (2013-2022).

Education:

Residence in Albert Einstein Israeli Hospital (HIAE) Sao Paulo/SP (2009-2012), University of Sao Paulo, Medical School. Expertise in the Intensive Care of Burns at the Intensive Care Unit of hospital - Central Institute of Clinics Hospital of Medical School of University of Sao Paulo, since 2013. "Anesthesiology, Surgical Sciences and Perioperative Medicine" of Medical School of University of Sao Paulo (2021).

Research Experience:

Intensive on duty in ICU- Division of Plastic Surgery and Burns of Surgery Department of Medical School of University of São Paulo (2013-23).

Professional Experience:

Since 2013 actively participated in the implementation of the PK/PD approach based on serum levels of antimicrobials in the burn ICU for therapeutic drug monitoring related to antimicrobial effectiveness and safety. Individualization of therapy in ICU critically ill pediatric and adult burn patients started in 2007 at the laboratory of the Clinical Pharmacokinetics Center – University of Sao Paulo.

Professional Appointments:

Research fellow of team on Clinical Pharmacokinetics in tertiary public hospital of Medical School, University of Sao Paulo (2013-2023).

Honors: Scientific Awards from the Last 3 Years:

III International Meeting Sobratafe on Advanced Wound Care, Sobratfe 2019, Maksoud Plaza Hotel São Paulo, SP. Awards at The Event Promoter: Brazilian Society for Advanced Treatment of Complex Wounds (Sobratafe)

Award 1: Vancomycin dose adjustment done in areal time for target attainment in critically ill septic pediatric burns versus non-burns against Staphylococcus spp MIC 1 mg/L. Camila Salinas Guillaux, Ronaldo Morales Júnior, Vedilaine Aparecida Bueno da Silva Macedo, Frederico Ribeiro Pires, Elson Mendes da Silva Júnior, João Manoel da Silva Júnior, Edvaldo Vieira de Campos, Amanda Maria Ribas Rosa de Oliveira, Aline Sandré Gomides, Gabriela Aparecida Ferreira, Thiago C. Oliveira, Silvia Regina Cavani Jorge Santos, David de Souza Gomez.

Award 2: Meropenem effectiveness in critically ill burn patients during the septic shock by pharmacokinetics assessment based on drug serum measurements done in a real time. Claudia Garcia Messiano, Elson Mendes da Silva Júnior, João Manoel da Silva Júnior, Edvaldo Vieira de Campos, Amanda Maria Ribas Rosa de Oliveira, Aline S. Gomides, Gabriela A. Ferreira, Thiago C. Oliveira, Silvia Regina Cavani Jorge Santos, David de Souza Gomez.

National Meeting Sobratafe on Advanced Wound Care, Sobratfe 2020, Maksoud Plaza Hotel São Paulo, SP. Four Awards at The Event. Promoter:

Brazilian Society for Advanced Treatment of Complex Wounds Dr Walter Soares Pinto Prize.

"Clinical Protocol" Award 1: Is Meropenem effectiveness in ICU burn patients during SIRS at the earlier period of septic shock impacted by pharmacokinetic changes? Adolescents versus Young adults. Thais V. de Camargo, Edvaldo Vieira de Campos, Amanda Maria Ribas Rosa de Oliveira, João Manoel da Silva Júnior, Elson Mendes da Silva Júnior, Aline S. Gomides, Gabriela A. Ferreira, Thiago C. Oliveira, Vera Lúcia Lanchote, Silvia Regina Cavani Jorge Santos, David de Souza Gomez.

"Clinical Protocol" Award 2: Extended infusion improves beta-lactams effectiveness in septic burn patients with normal renal function against P aeruginosa intermediate susceptibility. Vanessa Kasubeck de Souza, Karina B Vianna, João Manoel da Silva Júnior, Amanda Maria Ribas Rosa de Oliveira, Edvaldo Vieira de Campos, Elson Mendes da Silva Júnior, Aline S. Gomides, Gabriela A. Ferreira, Thiago C. Oliveira, Vera Lúcia Lanchote, Silvia Regina Cavani Jorge Santos, David de Souza Gomez.

Apfid-Prize (Asian Pacific Foundation For Infectious Diseases) "2021 Outstanding Poster Award" PhD. student Leonard de Vinci Kanda Kupa, R. *Morales, Jr*, Edvaldo Vieira de Campos, João Manoel da Silva Júnior, Elson Mendes da Silva Junior, David de Souza Gomez & Silvia R. C. J. Santos. https://doi.org/10.1016/j.ijantimicag.2021.106421.10.

Research team member since 2013

Publications from the Last 3 Years

1) Souza, V. K., Silva Jr, J. M., Junior, E. M. S., Ferreira, G. A., Silva Filho, C. R., Santos, V. J., Gomez, D. S. & Santos, S. R. C. J (2021). Estudo comparativo da efetividade da piperacilina-tazobactana após infusão intermitente versus estendida em pacientes sépticos grandes queimados pela abordagem farmacocinética-farmacodinâmica (PK/PD). *The Brazilian Journal of Infectious Diseases*, 25, 101387. https://doi.org/10.1016/j.bjid.2020.101387.
2) Souza V. K., Silva Jr J. M., Silva Jr E. M., Ferreira, G. A., Silva Filho, C. R., Santos, V. J., Rocha A., Lanchote, V. L., Gomez D. S., Santos S. R. C. J. Estudo comparativo da efetividade da piperacilina-tazobactana após infusão intermitente versus estendida em pacientes sépticos grandes queimados pela abordagem farmacocinética-farmacodinâmica (PK/PD)

The Brazilian Journal of Infectious Diseases, 26,101337. https://doi.org/10.1016/j.bjid.2021.101337.

3) Morales Junior, R. M., Kupa, L. D., V. K., de Oliveira, E. M., de Campos, E. V., da Silva Junior, J. M., Junior, E. M. S., Gomez, D. S. & Santos, S. R. C. J. (2022). Cobertura antimicrobiana da vancomicina contra patógenos gram-positivos em pacientes queimados. *The Brazilian Journal of Infectious Diseases,* 26, 101745. https://doi.org/10.1016/j.bjid.2021.101745.
4) Morales, Jr., R., Kupa, L. V. K., Santos, V. J., Romano, P., Oliveira, E. M., Campos, E. V, Silva Jr, J. M., Silva Jr, E. M., Gomides A. S., Gomez, D. S. & Santos, S. R. C. J. (2021). Vancomycin dose adjustment against Gram-positive MIC 2 mg/L strains in critically ill adult burn patients by pharmacokinetic-pharmacodynamic approach. In *International Journal of Antimicrobial Agents.* Elsevier. https://doi.org/10.1016/j.ijantimicag.2021.106421.8.
5) Morales, Jr, R., Kupa, L. V. K., Silva Jr, J. M., Silva Jr, E. M., Oliveira, A. M. R. R., Gomides, A. S., Ferreira G. A., Gomez, D. S. & Santos, S. R. C. J. (2021). Meropenem extended infusion against nosocomial MIC 4 mg/L strains to guarantee drug effectiveness in burn patients at the earlier period of septic shock. *International Journal of Antimicrobial Agents,* 58, 21002806. https://doi.org/10.1016/j.ijantimicag.2021.106421.9.
6) Morales, Jr, R., Kupa, L. V. K., Vianna, K. B., Garcia, M. S., Santos, V. J., Campos, E. V., Silva Jr, J. M., Silva Jr E. M., Oliveira, T. C., Gomez, D. S. & Santos, S. R. C. J. (2021). Meropenem extended infusion versus intermittent infusion against nosocomial MIC 4 mg/L strains to guarantee drug effectiveness by PK/PD approach in burn patients at the earlier period of septic shock. In *International Journal of Antimicrobial Agents.* Elsevier. https://doi.org/10.1016/j.ijantimicag.2021.106421.10.
7) Souza, V. K., Silva Jr E. M., Kaneko V. M., Silva Jr J. M., Gomides A. S., Ferreira G. A., Oliveira T. C., Gomez D. S., Santos, S. R. C. J. Extended two hours infusion improves piperacillin effectiveness against susceptible strains in septic burn patients. *Critical Care* (2019). v. 23(3): p 21-22, https://doi.org/10.1186/s13054-019-9. ISSN 1364-8535.
8) Messiano, C. G., Junior, R. M., Pereira, G. O., da Silva Junior, E. M., Gomez, D. S. & Santos, S. R. C. J. (2022). Therapeutic Target Attainment of 3-Hour Extended Infusion of Meropenem in Patients with Septic Burns. *Clinical Therapeutics*, 44(4), 624-629. http://dx.doi.org/10.1016/j.clinthera.2022.02.007.

9) de Camargo, T. V., Junior, E. M. S., Silva Jr, J. M., Gomez, D. S. & Santos, S. R. C. J. (2022). PK/PD approach to evaluate Meropenem effectiveness in critically ill burn adolescents versus young adults undergoing therapy of septic shock. *Pharm Pharmacol Int J*, 10(3), 79-85 (2022). https://doi.org/10.15406/ppij.2022.10.00368.
10) Santos, S. R. C. J., de Camargo T. V., Macedo V. A., Souza, V. K. & Gomez, D. S. Vancomycin and meropenem serum monitoring for target attainment by PK/PD approach as an effective tool in the battle against nosocomial pathogens in septic pediatric burn patients. *Pharm Pharmacol Int J.* 2022;10(3):87–95 (2022). https://doi.org/10.15406/ppij.2022.10.00369.
11) Santos, S. R. C. J., de Camargo T. V., Messiano C. G., Kupa L. V. K., Souza V. K., Ronaldo Morales Jr. R., Sanches Pinto D. C. S., Silva Junior E. M., Silva Junior J. M., Gomez D. S. Combating bacterial resistance to antimicrobials in severe septic ICU patients: importance of meropenem, piperacillin serum monitoring as a dose adjustment and duration of infusion strategies. *Pharm Pharmacol Int J.* 11(2):52–60 (2023). https://doi.org/10.15406/ppij.2023.11.00402.
12) Santos, S. R. C. J., de Camargo T. V., Messiano C. G., Souza V. K. Frederico Ribeiro Pires, Débora Cristina Sanches Pinto, Edvaldo Vieira de Campos, Elson Mendes da Silva Junior, João Manoel da Silva Junior, David de Souza Gomez. Combined vancomycin-meropenem therapy in pediatric major burns undergoing therapy of septic shock guided by cultures and pharmacokinetic-pharmacodynamics approach based on serum levels to combat bacterial resistance. *Pharm Pharmacol Int J.* 11(2):62–68 (2023). https://doi.org/10.15406/ppij.2023.11.00403.

David de Souza Gomez
(Professor of Plastic Surgery and Burns)
Leadership 2 of Research Team

ORCID https://orcid.org/0000-0003-4539-7315 Brazilian Platform of Research (CNPq).

Affiliation:
University of São Paulo, Division of Plastic Surgery & Burns of Clinics Hospital of Medical School.

Education:

University of São Paulo Medical School (1978); Residence in Plastic Surgery (four years), (1980-1984) MD Plastic Surgery and Burns (1985).

Research Experience:

Leadership of Research Team Group of Brazilian Council of Research (CNPq) Group of a team of 37 physicians, surgeons, intensivists, nurses, clinical pharmacists, also including the antimicrobial serum monitoring done by the laboratory team of hospital. (2007-2023).

Membership of international Society:

International Society of Burn Injuries (ISBI).

Professional Experience:

Director of Burns Unit HCFMUSP (20 beds) Intensive Care Unit and Infirmary, Surgical Center, and Emergency Room – Division of Plastic Surgery and Burns. Expertise of group at the last 20 years in the tertiary public hospital is focused on major burns surgical and clinical treatment of critically ill patients including bed side monitoring the septic shock caused by nosocomial bacteria. These patients are undergoing combined therapy of bacterial infections caused by Gram-positive and Gram-negative strains with antibiotics, or even combined with antifungal agents. Main goals of the team are to guarantee antimicrobial effectiveness and safety by individualization of therapy to reach desired outcome, despite extended length of stay of these patients in ICU.

Professional Appointments:

Physician Assistant of Professor (1985-1989).

Supervising Physician of the Plastic Surgery and Burns Division (1990-2008)

Director of Burn Unit (2008-2023), Chairman of Committees CCIH (1988-2023) of Central Institute of Clinics Hospital, Medical School, University of Sao Paulo, Sao Paulo/SP.

Brazilian Society for Advanced Treatment of Complex Wounds (Sobratafe) as Treasurer Director (1995-2023).

Honors:

Professor University of Sao Paulo, Fellowship of National Society Brazilian Society for Advanced Treatment of Complex Wounds.

Scientific Prizes last three years: National and international scientific events:

Brazilian Society of Infectiology (BSI) December/2020. The Brazilian Journal of Infectious Diseases, Volume 25, Supplement 1, January 2021 ISSN 1413-86-70. https://doi.org/10.1016/j.bjid.2020.101387.

"Comparative study of the effectiveness of Piperacillin-Tazobactam after intermittent versus extended infusion in severe burn septic patients by the pharmacokinetic-pharmacodynamic approach (PK/PD)." Vanessa Kasubeck de Souza Master of Sciences Student, dvisor S. R. C. J. Santos.

APFID-Prize (Asian Pacific Foundation for Infectious Diseases) "2021 Outstanding Poster Award" promoted by Asian Pacific Foundation for Infectious Diseases. Pos-doc Student Leonard de Vinci Kanda Kupa. https://doi.org/10.1016/j.ijantimicag.2021.106421.10.

III International Meeting Sobratafe on Advanced Wound Care, Sobratfe 2019, Promoted by Brazilian Society for Advanced Treatment of Complex Wounds.

Award 1: Meropenem effectiveness in critically ill burn patients during the septic shock by pharmacokinetics assessment based on drug serum measurements done in a real time. Claudia Garcia Messiano, Elson Mendes da Silva Júnior, João Manoel da Silva Júnior, Edvaldo Vieira de Campos, Amanda Maria Ribas Rosa de Oliveira, Aline S. Gomides, Gabriela A. Ferreira, Thiago C. Oliveira, Silvia Regina Cavani Jorge Santos, David de Souza Gomez.

Award 2: Vancomycin dose adjustment done in areal time for target attainment in critically ill septic pediatric burns versus non-burns against Staphylococcus spp MIC 1 mg/L. Camila Salinas Guillaux, Ronaldo Morales Júnior, Vedilaine Aparecida Bueno da Silva Macedo, Elson Mendes da Silva Júnior, João Manoel da Silva Júnior, Edvaldo Vieira de Campos, Amanda Maria Ribas Rosa de Oliveira, Aline Sandré Gomides, Gabriela Aparecida Ferreira, Thiago C. Oliveira, Silvia Regina Cavani Jorge Santos, David de Souza Gomez.

National Meeting Sobratafe on Advanced Wound Care, Sobratfe 2020. Promoted by the Brazilian Society for Advanced Treatment of Complex Wounds

Dr Walter Soares Pinto Prize

Award 1: Is Meropenem effectiveness in ICU burn patients during SIRS at the earlier period of septic shock impacted by pharmacokinetic changes? Adolescents versus Young adults. Thais V. de Camargo, Edvaldo Vieira de Campos, Amanda Maria Ribas Rosa de Oliveira, João Manoel da Silva Júnior, Elson Mendes da Silva Júnior, Aline S. Gomides, Gabriela A. Ferreira, Thiago C. Oliveira, Vera Lúcia Lanchote, Silvia Regina Cavani Jorge Santos, David de Souza Gomez.

Award 2: Extended infusion improves beta-lactams effectiveness in septic burn patients with normal renal function against P aeruginosa intermediate susceptibility. Vanessa Kasubeck de Souza, Karina B Vianna, João Manoel da Silva Júnior, Amanda Maria Ribas Rosa de Oliveira, Edvaldo Vieira de Campos, Elson Mendes da Silva Júnior, Aline S. Gomides, Gabriela A. Ferreira, Thiago C. Oliveira, Vera Lúcia Lanchote, Silvia Regina Cavani Jorge Santos, David de Souza Gomez.

Award 3: Pharmacokinetic-pharmacodynamic approach permits vancomycin dose adjustment for effectiveness against infection caused by Gram-positive pathogens MIC > 1 mg/L in ICU paediatric burn patients. Ronaldo Morales Júnior, Thais Vieira de Camargo, Vedilaine Aparecida Bueno da Silva Macedo, Edvaldo Vieira de Campos, João Manoel da Silva Júnior, Nilo José Coelho Duarte, Silvia Regina Cavani Jorge Santos, David de Souza Gomez.

Award 4: Is beta-lactams effectiveness impacted by the duration of drug infusion in critically ill septic patients. A systematic review of clinical PK-studies based on PK/PD approach. Karina B Vianna, João Manoel da Silva Júnior, Elson Mendes da Silva Júnior, David de Souza Gomez, Silvia Regina Cavani Jorge Santos.

Publications at the last three years:

1) Otofuji, G., Silva Jr, J. M., Silva Junior, EM., Gomides, A. S., Rocha, A., Gomez, D. S. Lanchote V. L. & Santos, S. R. C. J. (2021). The change in medical behavior in the prescription of antimicrobials guided by cultures, biomarkers and PK/PD approach of meropenem in a septic patient with severe burns with infection caused by amikacin-resistant p aeruginosa. *The Brazilian Journal of Infectious Diseases,* 25, 101409. https://doi.org/10.1016/j.bjid.2020.101409.
2) Matsuno, V. K., Junior, J. M. S., Gomides, A. S., Silva Filho, C. R., Santos, V. J., Rocha, A., Gomez, D. S. & Santos, S. R. C. J. (2021).

Fluconazole PK/PD approach to ensure effectiveness of antifungal therapy in critically ill, major burn septic patients with systemic Candida glabrata infection (CIM 32 mg/L). *The Brazilian Journal of Infectious Diseases,* 25, 101438. https://doi.org/10.1016/j.bjid.2020.101438.

3) dos Santos, M. S., Silva Jr, J. M., Silva Junior, E. M., Gomides, A. S., Rocha A., Duarte, A. J. S., Gomez, D. S. Lanchote V. L. & Santos, S. R. C. J (2021). Effectiveness of antimicrobial therapy in severe burn septic patient with renal dysfunction and severe ocular injury in intensive therapy with vancomycin and piperacillin-tazobactan through the pharmacokinetic-pharmacodynamic approach. *The Brazilian Journal of Infectious Diseases,* 25, 101337. Maria Severina dos Santos Bachelor of Pharmaceutical Sciences. https://doi.org/10.1016/j.bjid.2020.101337.
4) Souza, V. K., Silva Jr, J. M., Junior, E. M. S., Ferreira, G. A., Silva Filho, C. R., Santos, V. J., Gomez, D. S. & Santos, S. R. C. J (2021). Comparative study of the effectiveness of piperacillin-tazobactan after intermittent versus extended infusion in severe burn septic patients by the pharmacokinetic-pharmacodynamic approach (PK/PD). *The Brazilian Journal of Infectious Diseases*, 25, 101387. https://doi.org/10.1016/j.bjid.2020.101387.
5) De Camargo, T. V., da Silva Jr, J. M. M., da Silva Junior, E. M., da Silva Filho, C. R., Santos, V. J., Oliveira, T. C., Gomez, D. S. & Santos, S. R. C. J. (2021). Comparative study of the effectiveness of meropenem in septic burn patients. Teenagers versus young adults. *The Brazilian Journal of Infectious Diseases,* 25, 101388. https://doi.org/10.1016/j.bjid.2020.101388.
6) Romano, P., Duarte, N. J. C., Silva Junior, J. M., Silva Junior, E. M., Sandre, A., Rocha, Duarte, A. J. S., A Gomez, D. S., Lanchote V. L. & Santos, S. R. C. J. (2021). Effectiveness of antimicrobial therapy in a large burnt elderly septic patient in intensive care with vancomycin-meropenem through the pharmacokinetic-pharmacodynamic approach. *The Brazilian J. Inf. Diseases,* 25, 101403. https://doi.org/10.1016/j.bjid.2020.101403.
7) Vianna, K. B., da Silva Jr, J. M., da Silva Jr, É. M., Oliveira, T. C., Silva Filho, C. R., Santos, V. J., Rocha A, Lanchote, V. L., Gomez, D. S. & Santos, S. R. C. J. (2021). Efficacy of meropenem after rapid versus extended infusion in septic burn patients by pharmacokinetic/pharmacodynamic approach (PK/PD). *The Brazilian Journal of Infectious Diseases,* 25, 101391. https://doi.org/10.1016/j.bjid.2020.101391.

8) Souza, V. K., Silva Jr J. M., Silva Jr E. M., Ferreira, G. A., Silva Filho, C. R., Santos, V. J., Rocha A., Lanchote, V. L., Gomez D. S., Santos S. R. C. J. Comparative study of the effectiveness of piperacillin-tazobactan after intermittent versus extended infusion in severe burn septic patients by the pharmacokinetic-pharmacodynamic approach (PK/PD) *The Brazilian Journal of Infectious Diseases,* 26,101337. https://doi.org/10.1016/j.bjid.2021.101337.
9) Otofuji, G., Morales Junior, R, Silva Junior, J. M., Silva Junior E. M., Oliveira, A. M. R. R., Oliveira, E. M., Ferreira G. A., Oliveira T. C., Gomez, D. S & Santos, S. R. (2022). PK/PD approach on the effectiveness of piperacillin-vancomycin in septic burn patients on an empiric regimen of the recommended dose in renal failure. *The Brazilian Journal of Infectious Diseases,* 26, 101737. https://doi.org/10.1016/j.bjid.2021.101737.
10) De Camargo, T. C., Kupa, L. V. K., Silva Junior, J. M., Silva Junior, E. M, Oliveira, A. M. R. R., Oliveira, E. M., Gomides A. S., Ferreira G. A., Oliveira T. C., Gomez, D. S. & Santos, S. R. (2022). Alteration of clinical management in critically ill septic patients with burns by adjusting the dose of vancomycin against gram-positive strains CIM 2 mg/L by the pharmacokinetic-pharmacodynamic approach. *Braz. Journal of Infectious Diseases*, 26, 101738. https://doi.org/10.1016/j.bjid.2021.101738.
11) De Camargo, T. V., da Silva Jr, J. M. M., da Silva Junior, E. M., da Silva Filho, C. R., Santos, V. J., Oliveira, T. C., Gomez, D. S. & Santos, S. R. C. J. Comparative study of the effectiveness of meropenem in septic burn patients. Teenagers versus young adults. *The Brazilian Journal of Infectious Diseases,* 25, 101388 (2021). https://doi.org/10.1016/j.bjid.2020.101388.
12) Morales Junior, R. M., Kupa, L. D., V. K., de Oliveira, E. M., de Campos, E. V., da Silva Junior, J. M., Junior, E. M. S., Gomez, D. S. & Santos, S. R. C. J. (2022). Antimicrobial coverage of vancomycin against gram-positive pathogens in burn patients. *The Brazilian Journal of Infectious Diseases,* 26, 101745. https://doi.org/10.1016/j.bjid.2021.101745.
13) Messiano, C. G., Silva Junior, E. M., Silva Junior, J. M., Gomides, A. S., Ferreira, G. A., de Oliveira, T. C., David S. Gomez, & Santos, S. R. (2022). Comparison of the effectiveness of meropenem in the early versus late phases of septic shock using the pharmacokinetic-pharmacodynamic approach in burn patients. *The Brazilian Journal of Infectious Diseases,* 26, 101746. https://doi.org/10.1016/j.bjid.2021.101746.

14) Romano, P., Ebner, P. A., dos Santos, M. S., Garcia, M. S., Kupa, L. D. V. K., Duarte, N. J. C., Gomez, D. S. & Santos, S. R. (2022). Piperacillin-meropenem serum monitoring by LCMS/MS to evaluate the effectiveness of the empirical dose regimen and the extended infusion strategy in septic burn patients through the pharmacokinetic-pharmacodynamic approach. *The Brazilian Journal of Infectious Diseases,* 26, 101756. https://doi.org/10.1016/j.bjid.2021.101756.
15) De Oliveira, E. M., de Almeida, J. P., Silva Junior, J. M., de Camargo, T. V., Sabanai, A. H., Siqueira, R. S., Gomez, D. S. & Santos, S. R. C. J. (2022). Optimization of vancomycin therapy in critically ill oncology patients in intensive care against gram-positive patients up to MIC 2 mg/L based on the pharmacokinetic-pharmacodynamic approach (PK/PD). *The Brazilian Journal of Infectious Diseases,* 26, 101759. https://doi.org/10.1016/j.bjid.2021.101759.
16) Morales, Jr., R., Kupa, L. V. K., Santos, V. J., Romano, P., Oliveira, E. M., Campos, E. V, Silva Jr, J. M., Silva Jr, E. M., Gomides A. S., Gomez, D. S. & Santos, S. R. C. J. (2021). Vancomycin dose adjustment against Gram-positive MIC 2 mg/L strains in critically ill adult burn patients by pharmacokinetic-pharmacodynamic approach. In *International Journal of Antimicrobial Agents.* Elsevier. https://doi.org/10.1016/j.ijantimicag.2021.106421.8.
17) Morales, Jr, R., Kupa, L. V. K., Silva Jr, J. M., Silva Jr, E. M., Oliveira, A. M. R. R., Gomides, A. S., Ferreira G. A., Gomez, D. S. & Santos, S. R. C. J. (2021). Meropenem extended infusion against nosocomial MIC 4 mg/L strains to guarantee drug effectiveness in burn patients at the earlier period of septic shock. *International Journal of Antimicrobial Agents,* 58, 21002806. https://doi.org/10.1016/j.ijantimicag.2021.106421.9.
18) Morales, Jr, R., Kupa, L. V. K., Vianna, K. B., Garcia, M. S., Santos, V. J., Campos, E. V., Silva Jr, J. M., Silva Jr E. M., Oliveira, T. C., Gomez, D. S. & Santos, S. R. C. J. (2021). Meropenem extended infusion versus intermittent infusion against nosocomial MIC 4 mg/L strains to guarantee drug effectiveness by PK/PD approach in burn patients at the earlier period of septic shock. In *International Journal of Antimicrobial Agents*. Elsevier. https://doi.org/10.1016/j.ijantimicag.2021.106421.10.
19) Morales, Jr, R., Macedo, V. A. S., Pires, F. R., Campos, E. V., Ebner, P. A. R., Romano, P., Maria Severina dos Santos, Duarte, N. J., Gomez, D. S. & Santos, S. R. C. J. Vancomycin target attainment in critically ill septic pediatric burn versus non-burn patients against Staphylococcus spp.

International Journal of Antimicrobial Agents, 58, 2100282. https://doi.org/10.1016/j.ijantimicag.2021.106421.11.

20) Morales, Jr, R., Chalom, M. Y., Romano, P., Ebner, P. A. R., Santos, M. S., Kupa, L. V. K., Sumita, N. M., Duarte N. J. C., Duarte, A. J. S., Gomez, D. S. & Santos, S. R. J. C. (2021). Meropenem 3 hrs-extended infusion improves effectiveness in critically ill burn patients at the earlier period of septic shock against P. aeruginosa intermediate susceptibility MIC 4-8 mg/L In *International Journal of Antimicrobial Agents*. Elsevier. https://doi.org/10.1016/j.ijantimicag.2021.106421.12.
21) Morales, Jr, R., Chalom, M. Y., Romano, P., Ebner, P. A. R., Santos, M. S., Kupa, L. V. K., Sumita, N. M., Duarte N. J. C., Duarte, A. J. S., Gomez, D. S. & Santos, S. R. J. C. (2021). Piperacillin/tazobactam effectiveness in septic burn patients requiring vasopressors by applying pharmacokinetics-pharmacodynamics approach. In *International Journal of Antimicrobial Agents*. Elsevier. https://doi.org/10.1016/j.ijantimicag.2021.106421.13.
22) Morales, Jr, R., Chalom, M. Y., Romano, P., Ebner, P. A. R., Santos, M. S., Garcia, M. S., Kupa, L. V. K., Duarte N. J. C., Duarte, A. J. S., Gomez, D. S. & Santos, S. R. J. C. (2021). Extended infusion improves meropenem and piperacillin/tazobactam effectiveness in septic burn patients with normal renal by applying pharmacokinetics-pharmacodynamics approach. In *International Journal of Antimicrobial Agents*. Elsevier. https://doi.org/10.1016/j.ijantimicag.2021.106421.14.
23) Souza, V. K., Silva Jr E. M., Kaneko V. M., Silva Jr J. M., Gomides A. S., Ferreira G. A., Oliveira T. C., Gomez D. S., Santos, S. R. C. J. Extended two hours infusion improves piperacillin effectiveness against susceptible strains in septic burn patients. *Critical Care* (2019). v.23(3): p 21-22, https://doi.org/10.1186/s13054-019-9. ISSN 1364-8535.
24) Messiano, C. G., Junior, R. M., Pereira, G. O., da Silva Junior, E. M., Gomez, D. S. & Santos, S. R. C. J. (2022). Therapeutic Target Attainment of 3-Hour Extended Infusion of Meropenem in Patients with Septic Burns. *Clinical Therapeutics*, 44(4), 624-629. http://dx.doi.org/10.1016/j.clinthera.2022.02.007.
25) De Camargo, T. V., Junior, E. M. S., Silva Jr, J. M., Gomez, D. S. & Santos, S. R. C. J. (2022). PK/PD approach to evaluate Meropenem effectiveness in critically ill burn adolescents versus young adults undergoing therapy of septic shock. *Pharm Pharmacol Int J*, 10(3), 79-85 (2022). https://doi.org/10.15406/ppij.2022.10.00368.

26) Santos, S. R. C. J., de Camargo T. V., Macedo V. A., Souza, V. K. & Gomez, D. S. Vancomycin and meropenem serum monitoring for target attainment by PK/PD approach as an effective tool in the battle against nosocomial pathogens in septic pediatric burn patients. *Pharm Pharmacol Int J.* 2022;10(3):87–95 (2022). https://doi.org/10.15406/ppij.2022.10.00369.
27) Santos, S. R. C. J., de Camargo T. V., Messiano C. G., Kupa L. V. K., Souza V. K., Ronaldo Morales Jr. R., Sanches Pinto D. C. S., Silva Junior E. M., Silva Junior J. M., Gomez D. S. Combating bacterial resistance to antimicrobials in severe septic ICU patients: importance of meropenem, piperacillin serum monitoring as a dose adjustment and duration of infusion strategies. *Pharm Pharmacol Int J.* 11(2):52–60 (2023a). https://doi.org/10.15406/ppij.2023.11.00402.
28) Santos, S. R. C. J., de Camargo T. V., Messiano C. G., Souza V. K. Frederico Ribeiro Pires, Débora Cristina Sanches Pinto, Edvaldo Vieira de Campos, Elson Mendes da Silva Junior, João Manoel da Silva Junior, David de Souza Gomez. Combined vancomycin-meropenem therapy in pediatric major burns undergoing therapy of septic shock guided by cultures and pharmacokinetic-pharmacodynamics approach based on serum levels to combat bacterial resistance. *Pharm Pharmacol Int J.* 11(2):62–68 (2023b). https://doi.org/10.15406/ppij.2023.11.00403.

Valentina Porta
Associate Professor

ORCID https://orcid.org/0000-0003-4103-6758 Brazilian Platform of Research (CNPq).

Affiliation:
University of São Paulo, School of Pharmaceutical Sciences. (1996-2023).

Education:
University of São Paulo Pharmacist- Biochemist (1989), Master of Sciences on Pharmacy (1993); Doctor of Pharmacy (PhD 1999). Post doctoral studies in Université d'Auvergne Equipe de Recherche Technologique Conception, Ingénierie et Développment de l'Aliment et du Médicament (ERT

CIDAM) Research International Biopharmaceuticals Laboratories (France, 2016).

Research Experience:

Bioequivalence, drug dissolution, in vitro-in vivo correlations, chromatographic bioanalytical techniques by application of liquid chromatography for dosage of drugs, Department of Pharmacy, Faculty of Pharmaceutical Sciences.

Membership of National Society:

Brazilian Society on Clinical Pharmacy.

Leadership of Research "Biopharmaceuticals" Advisor of Undergraduate Students since 2001, Department of Pharmacy, School of Pharmaceutical Sciences.

Professional Experience:

Bioequivalence, drug dissolution, in vitro-in vivo correlations, chromatographic bioanalytical techniques by application of liquid chromatography for dosage of drugs, Department of Pharmacy, Faculty of Pharmaceutical Sciences.

Professional Contests of the Faculty of Pharmaceutical Sciences of the University of São Paulo

1. Department of Pharmacy as Professor's Assistant in several disciplines: Biopharmaceuticals, Clinical Pharmacy, Hospital Pharmacy, Pharmacoeconomics, Patient Safety (1996-2023).
2. University of São Paulo, Faculty of Pharmaceutical Sciences: Lecturer (1996-2013).
3. University of São Paulo, Faculty of Pharmaceutical Sciences: Associate Professor (2013-2023)
4. Supervision of research projects related to the Bioavailability Studies Center of the Faculty of Pharmaceutical Sciences (2000-2008).
5. Ethics Committee, Graduate Studies Committee, Research Committee (2010-2023).

Honors:

Associate Professor at the University of São Paulo

Director of Pharmacy & Clinical Pharmacy Division University Hospital of University of Sao Paulo, campus São Paulo/SP.

Beatriz Aparecida Passos Bismara Paranhos

ORCID https://orcid.org/0000-0002-8777-5778 Brazilian Platform of Research (CNPq).

Affiliation:

University of São Paulo, School of Pharmaceutical Sciences (since 2008).

Education:

Centro Universitario Herminio Ometto de Araras, School of Pharmacy: Pharmacist (2001), Master of Pharmacology, Faculty of Medical Sciences of State University of Campinas, FCM-UNICAMP (2006), Toxicological Analysis expert (2008) with experience in Pharmacy and emphasis on Molecular Biology and Toxicological Analysis. Master Business Administration MBA "Health Management," two years in the Faculty Getulio Vargas (2006).

Research Experience:

Drug measurements in several biological matrices by GC, GC-MS technique and UHPLCMS/MS. Multiuser Laboratory of Mass Spectrometer, Department of Clinical & Toxicological Analysis. Multiuser Laboratory of Mass Spectrometry (LMEM).

Assistant of Research related to Clinical Toxicology.

Supervision of Graduate and Undergraduate Students in the Multi-User Laboratory of Mass Spectrometry, Department of Clinical & Toxicological Analysis since 2008.

Membership of National or International Society:

The International Association of Forensic Toxicologists (TIAFT) since 2022.

Professional Experience:

Toxicological Analysis Expertise since 2008 on chromatographic bioanalytical techniques by applying gas and liquid chromatography and mass spectrometry for drug measurements in biological samples in the Multi-User Laboratory of Analytical Center, Department of Clinical & Toxicological Analysis (2008-2023).

Professional Appointments at the School of Pharmaceutical Sciences at the University of Sao Paulo.

Clinical & Toxicological Department: Supervision of Graduate Studies students in Toxicological Analysis since 2008.

Honors:

Member of Research-Team from Clinical Pharmacokinetics Group since 2022.

Publications at the last three years:

1) Durão A. C. C. S., Brandão W. N., Bruno V., Spelta L. E. W., Duro S. O., dos Santos N. B., Paranhos B. A. P. B., Zanluqui N. G., Yonamine M., Peron J. P. S., Munhoz C. D., Marcourakis T. In Utero Exposure to Environmental Tobacco Smoke Increases Neuroinflammation in Offspring. *Front. Toxicol.* 17 January (2022). https://doi.org/10.3389/ftox.2021.802542.
2) Tomaz P. R. X., Gonçalves T. S., Santos J. R., Scholz J., Abe T. O., Gaya P. V., Figueiredo E. C., de Faria H. D., Martins I., Fonseca Pego A. M. F., Bismara B. A., Yonamine M., Pereira A. C., Santos P. C. J. L. Evaluation of the nicotine metabolite ratio in smoking patients treated with varenicline and bupropion. *Front Toxicol.* 22 July (2022). https://doi.org/10.3389/fphar.2022.900112.
3) Bouso J. C., Jiménez-Garrido D., Ona G., Woznica D., dos Santos R. G., Hallak J. E. C., Paranhos B. A. P. B., Mendes F. A., Yonamine M., Alcázar-Córcoles M. Á., Farré M. Quality of life, mental health, personality, and patterns of use in self-medicated cannabis users with chronic diseases: a 12-month longitudinal study. *Phytother Res.* 34:1670-1677(2020). doi: https://doi.org/10.1002/ptr.6639 [PubMed] Citation in doi: https://doi.org/10.18176/resp.00044.

4) Jiménez-Vinaja R., M. Imbernón, A. Parés-Bayerri, G. Ona, O. Parés, J. C. Bouso. Jailhouse rock: Health status, patterns of use, and subjective benefits of cannabis use in a sample of inmates in Spanish prisons. *Rev Esp Sanid Penit.* 2022 Jan-Apr; 24(1): 9–14. Published online 2022 Mar 31. PMCID: PMC9017606. PMID: 35411912. https://doi.org/10.1002/ptr.6639.

Mauricio Yonamine
Associate Professor

ORCID https://orcid.org/0000-0002-4615-1088 Brazilian Platform of Research (CNPq).

Affiliation:
University of São Paulo, School of Pharmaceutical Sciences. (2005-2023).

Education:
University of São Paulo: Pharmacist (1996), Master of Toxicology (2000); Doctor of Toxicology (2004).

Research Experience:
Leadership of Research "Clinical Toxicology" Advisor of Graduated Students in the Multi-User Laboratory of Analytical Center, Department of Clinical & Toxicological Analysis of School of Pharmaceutical Sciences.

Professional Experience at the last 19 years in toxicological analysis. Expertise in development and application of toxicological analysis, with emphasis on drugs of abuse, working mainly on the following subjects: sample preparation techniques (SPE, SPME, LPME, DLLME, SHS, accelerated solvent extraction), chromatographic techniques and mass spectrometry, alternative biological matrices (hair, sweat, oral fluid, meconium, umbilical cord), mainly in aspects related to Social/Forensic Toxicology.

Professional Appointments at the School of Pharmaceutical Sciences, University of Sao Paulo:

1. Clinical & Toxicological Department as Assistant of Professor on several disciplines (Toxicology, Clinical Toxicology, Therapeutic

Drug Monitoring, Drug abuse) for Undergraduate students and Graduate Studies students (2004-2010).

2. University of Sao Paulo, School of Pharmaceutical Sciences: Lecturer (2011).
3. University of Sao Paulo, School of Pharmaceutical Sciences: Associate Professor (2012-2023)
4. President of Committees at the School of Pharmaceutical Sciences Ethics, Graduate Studies (2019-2022)
5. President of the Multiuser Laboratory of Mass Spectrometry (LMEM), linked to USPMulti.

Honors:

Associate Professor of University of Sao Paulo; Fellowship of International Toxicology Societies.

Membership of National and International Society:

International Association of Forensic Toxicologists (TIAFT), Brazilian Society of Toxicology SBTOx. Vice-President of the Brazilian Society of Toxicology SBTOx (2016-2019).

Publications at the last three years:

1) Lima M. B. A., Yonamine M. Counterfeit medicines: relevance, consequences, and strategies to combat the global crisis. *Braz J Pharm Sci.* (2023) v.59: e20402. http://dx.doi.org/10.1590/s2175-9790 2023e20402.
2) Tomaz P. R. X., Gonçalves T. S., Santos J. R., Scholz J., Abe T. O., Gaya P. V., Figueiredo E. C., de Faria H. D., Martins I., Fonseca Pego A. M. F., Bismara B. A. P., Yonamine M., Pereira A. C., Santos P. C. J. L. Evaluation of the nicotine metabolite ratio in smoking patients treated with varenicline and bupropion. *Front Toxicol.* 22 July (2022). https://doi.org/10.3389/fphar.2022.900112.
3) Durão A. C. C. S., Brandão W. N., Bruno V., Spelta L. E. W., Duro S. O., dos Santos N. B., Paranhos B. A. P. B., Zanluqui N. G., Yonamine M., Peron J. P. S., Munhoz C. D., Marcourakis T. In Utero Exposure to Environmental Tobacco Smoke Increases Neuroinflammation in Offspring. Front. Toxicol. 17 January 2022 Sec. *Neurotoxicology.* v.3 (2021). https://doi.org/10.3389/ftox.2021.802542.

4) Rodrigues T. B., Souza M. P., Barbosa L. M., Ponce J. C., Neves Júnior L. F., Yonamine M., Cost J. L. Synthetic cannabinoid receptor agonists profile in infused papers seized in Brazilian prisons. *Forensic Toxicology,* 12 June © Japanese Association of Forensic Toxicology (2021). https://doi.org/10.1007/s11419-021-00586-7.
5) Almeida C. A. F., Pereira-Junior A. A., Rangel J. G., Pereira B. P., Costa K. C. M., Bruno V., Silveira G. O., Cerona C. S., Yonamine M., Camarinic R., Garcia R. C. T., Marcourakis T., Torres H. Ayahuasca, a psychedelic beverage, modulates neuroplasticity induced by ethanol in mice. *Behavioral Brain Res.* v. 416, 7 January 2022, 113546. https://doi.org/10.1016/j.bbr.2021.113546.
6) Fabris A. L., Lanaro R., Costa J. L., Yonamine M. Development of a Dispersive Liquid–Liquid Microextraction for Synthetic Cathinone's in Biological Fluids Based on Principles of Green Analytical Toxicology. *J Analyt Toxicol.* (2023) v47(4):353–365. May 2023. https://doi.org/10.1093/jat/bkad003.
7) dos Santos R. G., Rocha J. M., Rossi G. N., Osório F. L., Ona G., Bouso J. C., Silveira G. O., Yonamine M., Marchioni C., Crevelin E. J., Queiroz M. E., Crippa J. A., Hallak J. E. C. Effects of ayahuasca on the endocannabinoid system of healthy volunteers and in volunteers with social anxiety disorder: Results from two pilot, proof-of-concept, randomized, placebo-controlled trials. *Human Pharmacol.* 2 February (2022). https://doi.org/10.1002/hup.2834.
8) Daneluz D. M., Sohn J. M. B., Silveira G. O., Yonamine Y., Aparecida Stern C. A. Evidence on the impairing effects of Ayahuasca on fear memory reconsolidation. *Psychopharmacology volume* 239, pages 3325–3336 (2022). Published: 07 September (2022). https://doi.org/10.1007/s00213-022-06217-2.
9) Rocha J. M., Rossi G. N., Osório, F. L., Saiz J. C. B., Silveira G. O., Yonamine M., Crevelin E. J., Queiroz M. E., Hallak, J. E. C., dos Santos R. G. Effects of Ayahuasca on Personality: Results of Two Randomized, Placebo-Controlled Trials in Healthy Volunteers. *Front Psychiatry* (2021)v.12:688439. Published online 2021 Aug 6. PMCID: PMC8377499. PMID: 34421675. https://doi.org/10.3389/fpsyt.2021.688439.
10) Souza F. R., dos Santos M. R., Rochitte C. E., dos Santos R. P., Jordão C. P., Leite I. S., da Fonseca G. W. P., Fonseca R. A., de Oliveira T. F., Yonamine M., Pereira R. M. R., Negrão C. E., de Nazaré Nunes M. J. A. Decreased Native T1 Values and Impaired Myocardial Contractility in

Anabolic Steroid Users. *Int J Sports Med.* 43(02): 183-191 (2022). https://doi.org/10.1055/a-1518-7953.

11) Zucoloto A. D., Eller S., de Oliveira T. F., Wagner G. A., Fruchtengarten L. V. G., de Oliveira C. D. R., Yonamine M. Relationship between cocaine and cocaethylene blood concentration with the severity of clinical manifestations. *The American Journal of Emergency Medicine.* v. 50:404-408. December (2021). https://doi.org/10.1016/j.ajem.2021.08.057.
12) Fabris A. L., & Mauricio Yonamine. Dried matrix spots in forensic toxicology. *Bioanalysis.* (2021) v. 13(18). Published Online :23 Septembre 2021. https://doi.org/10.4155/bio-2021-0135.
13) Zucoloto A. D., Eller S., de Oliveira T. F., Wagner G. A., Fruchtengarten L. V. G., D. R. de Oliveira C. D. R., Yonamine M. Relationship between cocaine and cocaethylene blood concentration with the severity of clinical manifestations. *The American Journal of Emergency Medicine* (2021) v. 50: 404-408. December 2021. https://doi.org/10.1016/j.ajem.2021.08.057.
14) Xavier J., Farias C. P., Soares M. S. P., Silveira G. O., Spanevello R. M., Yonamine M., Gamaro G. D., de Carvalho H. W., Cognato G. P. Ayahuasca prevents oxidative stress in a rat model of depression elicited by unpredictable chronic mild stress. *Arc Clin Psychiatr* 48 (2). Mar-Apr 2021. https://doi.org/10.15761/0101-60830000000286.
15) Roveri F. L., Pego A. M. F., Eller S., de Oliveira T. F., Yonamine M. Evaluation of artificial drug incorporation into hair for the production of quality control samples. *Drug Analytical Research.* (2021) v5(2). https://doi.org/10.22456/2527-2616.118887.
16) Bouso J. C., Jiménez-Garrido D., Ona G., Woznica D., dos Santos R. G., Hallak J. E. C., Paranhos B. A. P. B., Mendes F. A., Yonamine M., Alcázar-Córcoles M. Á., Farré M. Quality of life, mental health, personality, and patterns of use in self-medicated cannabis users with chronic diseases: a 12-month longitudinal study. *Phytother Res.* (2020) v34:1670-1677. https://doi.org/10.1002/ptr.6639.
17) Silveira G. O., dos Santos, R. G., Lourenço F. L., Rossi G. N., Hallak J. E. C., Yonamine M. Stability Evaluation of DMT and Harmala Alkaloids in Ayahuasca Tea Samples. *Molecules.* (2020) v25(9), 2072. Published: 29 Apr. 2020. https://doi.org/10.3390/molecules25092072.
18) Linhares A. L. F. A. & Yonamine M. Analysis of biofluids by paper spray-MS in forensic toxicology. *Bioanalysis* (2020) v. 12(15). https://doi.org/10.4155/bio-2020-0160.

19) Pego A. M. F., V. Leyton, I. D., Miziara, R. H. Bortolin, R. C. C. Freitas, M. Hirata, P. R. X. Tomaz, J. R. Santos, P. C. J. L. Santos, M. Yonamine. SNPs from BCHE and DRD3 genes associated to cocaine abuse amongst violent individuals from Sao Paulo, Brazil. *Forensic Science International* (2020)v 317, Dec 2020, 11051. https://doi.org/10.1016/j.forsciint.2020.110511.
20) Silveira G. O., Lourenço F. L., Bruno V., Yonamine M. Fast Hollow Fiber Liquid-Phase Microextraction as a Greener Alternative for the Determination of N,N-Dimethyltryptamine and Harmala Alkaloids in Human Urine. *Front. Chem.* v. 8 (2020). Published 07 October. https://doi.org/10.3389/fchem.2020.558501.
21) Silveira G. O., Lourenço F. L., Pego A. M. F., dos Santos R. G., Rossi G. N., Hallak J. E. C., Yonamine M. Essential oil-based dispersive liquid-liquid microextraction for the determination of N,N-dimethyltryptamine and β-carbolines in human plasma: A novel solvent-free alternative. *Talanta.* v. 225, 1 April 2021, 121976. https://doi.org/10.1016/j.talanta.2020.121976.
22) Campos E. R., Eller S., Birk L., Coimbra M. A., Macedo S. M. D., Yonamine M., de Oliveira J., de Oliveira M. T. F. Analysis of seized stanozolol formulations in South Brazil by liquid chromatography coupled to quadrupole time-of-flight-mass spectrometry. *Drug Analytical Research.* (2020) 4(2):58-63. December 2020 License CC BY-NC-SA 4.0. https://doi.org/10.22456/2527-2616.108853.

Débora Cristina Sanches-Pinto
(Medical Doctor - Plastic Surgery and Burns)

ORCID https://orcid.org/0000-0001-9496-3024 *Brazilian Platform of Research (CNPq)*

Brazilian Platform ID lattes: http://lattes.cnpq.br/5441868541404798

Affiliation:

University of São Paulo, Division of Plastic Surgery & Burns at Hospital das Clínicas - University of São Paulo – Medical School.

Education:

University of São Paulo Medical School (1987); Residency in General Surgery (two years 1988 - 1989). Residency in Plastic Surgery and Burns (three years 1990 - 1992)

Business Address: Faria Lima Avenue, 1461 – 6th floor – Sao Paulo – SP – Brazil - Zip Code 04547-005. Hospital Address: Av. Dr. Eneas de Carvalho Aguiar, 155, Sao Paulo, SP, Brazil, Zip Code 05403-000.

Contacts:

e-mail: deborasanches29@uol.com.br.
website: http://www.dradeborasanchespinto.com.br/
Linkedin: Debora Cristina Sanches Pinto
Instagram/facebook: @dradeborasanchespinto

Research and Professional Experience:

1. Qualified plastic surgeon, working in aesthetic and reconstructive surgery, complex wounds, and burns since 1992.
2. Physician, University of São Paulo, Medical School- Brazil (1987).
3. Experience in university career, teaching undergraduate and graduate courses, in addition to intense participation as an invited speaker in national and international events since.
4. Worked in national and international research such as the use of bromelain, phase 3, in the beginning of its tests in Brazil.
5. Co-founder of the Brazilian Society for Advanced Wound Care (Sobratafe), currently acting as President of the Brazilian Society for Advanced Wound Care since 2013.
6. Master's Degree in Plastic Surgery and Burns from the School of Medicine of the University of São Paulo with a focus on Immunology of Burns and Reduction in the Late Mortality Rate of Severely Burned Patients using Cerium Nitrate Plus Silver Sulfadiazine (1999).
7. Finishing PhD in Medical Science with a focus on Chronic Wounds, comparing the use of Minced Skin Grafts with the use of mesh graft 1:3 (*in course*).
8. Executive MBA in Health at Faculty of Getúlio Vargas-Foundation and Hospital Administration at PROHASA (2010).

9. Ability to negotiate and discuss technical issues relevant to processes related to medical care, evaluation of new technologies and evidence-based cost-benefit studies.
10. Member of the Brazilian Society of Plastic Surgery since 1992.
11. Founding member of the Brazilian Society of Burns (1996).
12. Full member of International Societies: Latin American Federation of Burn, AAWC (American Association on Wound Care), WHS (Wound Healing Society), EBA (European Burn Association) and ISBI (International Society for Burn Injuries).

Professional Appointments:

1. Physician Assistant Professor of Plastic Surgery and Burns Division (1998 till now)
2. CEO of Cetrafe – Center of Advanced Wound Care (2017 till now)
3. Plastic and Reconstructive Surgery Chief of Santa Paula Hospital (2009 – 2018)
4. President of Brazilian Society on Advanced Wound Care (SOBRATAFE) (2005 till now)
5. President, organizer, and speaker at the following events

2008 – I Sobratafe AWC Symposium —Hospital do Servidor Público Estadual de São Paulo

2011 – I Sobratafe Brazilian Meeting and II Sobratafe AWC Symposium – Grand Hotel Senac - Águas de São Pedro – SP

2012 – I Sobratafe AWC Hands on Course Meeting – Hospital Santa Paula - SP

2013 – II Sobratafe Brazilian Meeting and III Sobratafe AWC Symposium - Grande Hotel Senac Campos do Jordão – SP

2014 – I Sobratafe AWC Hands on Course Meeting - Hospital Alemão Oswaldo Cruz

2014 – I Sobratafe International AWC Simposium and IV Sobratafe SP AWC Symposium

2015 – II Sobratafe International Meeting AWC / III Sobratafe Brazilian AWC Meeting

2016 – V Sobratafe Symposium on AWC

2017– IV Sobratafe Brazilian Meeting on Advanced Wound Care

2018 – I Sobfratafe South and Southeast Sobratafe AWC Symposium – Londrina - PR

2019 – III International Meeting Sobratafe on Advanced Wound Care
2020 – I Sobratafe AWC *online* Symposium
2021 – I Sobratafe Global Advanced Wound Care Meeting (SGAWCm - online)
2022 – VI Sobratafe Symposium on AWC in Sobenfee Meeting

Honors:

Member since 2022 of Research-Team of *Directory of the Research Group "Clinical Pharmacokinetics" of the Brazilian National Research Council, CNPq Platform.*

Center for Cultural and Business Integration of the State of São Paulo, recognition of the merits of honor, dignity, character, and civility, receiving the title of personality Brazil 500 years (December 20th, 2000).

Peru – Lima –SILAHUE Meeting- honored international guest.

Chile – Santiago – Chilean Burn Meeting - honored international guest (2006).

Colombia - Cartagena – Colombia Burn Meeting - honored international guest (2003).

Publications at the last three years

1) Lead author of the Tratado Latino-Americano de Feridas. Sanches -Pinto, Débora C.; Gomez, D. S.; Gutierrez, R. E. R.; Gemperli, R. *Tratado Latino-Americano de Feridas.* 1 ed. Rio de Janeiro: Guanabara Koogan, 2022. v. 1. 550p.
2) Sanches-Pinto, Débora C.; Gomez, D. S. Queimaduras. In: Sanches-Pinto DC, Gomez DS, Gutierrez RER, Gemperli R. (Org.). *Tratado Latino-Americano de Feridas.* 1 ed. Rio de Janeiro: Guanabara Koogan, 2022, v. 1, p. 105-118.2.
3) Sanches-Pinto D. C. Aspectos Imunológicos das Queimaduras. In: Sanches-Pinto DC, Gomez DS, Gutierrez RER, Gemperli R. (Org.). *Tratado Latino-Americano de Feridas.* 1 ed. Rio de Janeiro: Guanabara Koogan, 2022, v. 1, p. 119-124.
4) Sanches-Pinto D. C. História do Tratamento da Ferida. In: Sanches-Pinto DC, Gomez DS, Gutierrez RER, Gemperli R. (Org.). *Tratado Latino-Americano de Feridas.* 1 ed. Rio de Janeiro: Guanabara Koogan, 2022, v. 1, p. 1-10.
5) Sanches-Pinto D. C. Antimicrobianos Tópicos e Curativos com Propriedades Antinicrobianas. In: Sanches-Pinto DC, Gomez DS,

Gutierrez RER, Gemperli R. (Org.). *Tratado Latino-Americano de Feridas.* 1 ed. Rio de Janeiro: Guanabara Koogan, 2022, v. 1, p. 209-220.
6) Sanches-Pinto DC. Interdisciplinaridade. In: Sanches-Pinto DC, Gomez DS, Gutierrez RER, Gemperli R. (Org.). *Tratado Latino-Americano de Feridas.* 1 ed. Rio de Janeiro: Guanabara Koogan, 2022, v. 1, p. 477-480.
7) Sanches-Pinto, Débora C.; Eriksson, Elof; Gomez, David S.; Nunes, Maria P. T.; Gemperli, Rolf; Soriano, Francisco G. Minced skin grafts for chronic wounds compared to conventional mesh grafts. *Health Science Reports,* v. 6, p. e1353, 2023.
8) Sanches-Pinto, Débora C.; Mota, W. M.; Gomez, D. S.; Gemperli, R. 529 Collagen plus alginate wound dressing for the treatment of Split thickness Skin Graft Donor Sites. A Cost-effectiveness. *Journal of Burn Care & Research,* v. 41, p. S199-S200, 2020.
9) Sanches-Pinto, Débora C.; Mota, W. M.; Gomez, D. S.; Gemperli, R.. 735 Advantages of Biocellulose for Burn Injures. *Journal of Burn Care & Research,* v. 41, p. S199-S200, 2020.
10) Santos, Silvia R. C. J.; de Camargo, Thais Vieira; Garcia Messiano, Claudia; Kupa, Leonard de Vinci Kanda; Morales Jr, Ronaldo; de Souza, Vanessa Kazubeck; Pires, Frederico Ribeiro; Sanches-Pinto, Débora Cristina; de Campos, Edvaldo Vieira; da Silva Junior, Elson Mendes; da Silva Junior, João Manoel; Gomez, David de Souza. Combined vancomycin-meropenem therapy in pediatric major burns undergoing therapy of septic shock guided by cultures and pharmacokinetic-pharmacodynamics approach based on serum levels to combat bacterial resistance. *Pharmacy & Pharmacology International Journal,* v. 11(2), p. 62-68, 2023. https://doi.org/10.15406/ppij.2023.11.00402.
11) Santos, Silvia R. C. J.; de Camargo, Thais Vieira; Garcia Messiano, Claudia; Kupa, Leonard de Vinci Kanda; Morales Jr, Ronaldo; de Souza, Vanessa Kazubeck; Pires, Frederico Ribeiro; Sanches-Pinto, Débora Cristina; de Campos, Edvaldo Vieira; da Silva Junior, Elson Mendes; da Silva Junior, João Manoel; Gomez, David de Souza. Combating bacterial resistance to antimicrobials in severe septic ICU patients: importance of meropenem, piperacillin serum monitoring as a dose adjustment and duration of infusion strategies. *Pharm Pharmacol Int J.* 11(2):52–60 (2023). https://doi.org/10.15406/ppij.2023.11.00403.

Chapter 2

Preconcentration Strategies Coupled with Chromatographic Methods for the Determination of Organic Compounds

Raif İlktaç*
Application and Research Center for Testing and Analysis, University of Ege, Bornova, İzmir, Turkey

Abstract

Chromatographic methods are the most widely used analytical tools for the separation and detection of inorganic and organic compounds. In chromatographic methods, different separation systems can be combined with a wide range of detectors that allow analysis of a large number of compounds. Chromatographic methods have been used in different areas such as environmental, clinical, forensic and life sciences.

Preconcentration techniques aim to concentrate or enrich the target compounds. In this perspective, preconcentration techniques are used prior to the analytical determination procedures, both to separate and enrich the analyte and to minimize the matrix effects. Preconcentration techniques also improve the detection limits of the methods. The number of the novel materials and methods developed by researchers to use as a preconcentration tool is increasing rapidly. Present chapter focuses on the preconcentration strategies coupled with chromatographic methods used for the detection of various types of organic compounds.

Keywords: preconcentration, enrichment, separation, chromatography

---

* Corresponding Author's Email: raif.ilktac@ege.edu.tr.

In: Chromatographic Methods and Research
Editor: Raven W. Obrien
ISBN: 979-8-89113-193-4

Introduction

Chromatography is a powerful analytical technique which is used to separate, identify and determine the components of a mixture. The basic principle of the chromatography depends on the migration of the mixture through a stationary phase and a mobile phase. The stationary phase can be a solid or a liquid, while the mobile phase is usually a liquid or gas. Basically, the mixture is introduced into the stationary phase, which retains some components while allowing others to pass through it based on their physical and chemical properties. The retained components can then be eluted from the stationary phase by using a suitable solvent or by changing the conditions of the system (Coskun, 2016; Scott, 2003). Chromatography can be divided into six main groups due to their separation mechanism;

1. Gas Chromatography (GC): GC separates the volatile compounds by vaporizing them and passing them through a stationary phase, typically a packed column filled with a solid or a liquid. The separation is based on the differences in boiling points and affinities for the stationary phase.
2. Liquid Chromatography (LC): LC generally separates non-volatile compounds dissolved in a liquid mobile phase, which is passed through a stationary phase such as a column filled with a solid or a liquid. The separation is based on the differences in chemical properties such as; polarity, size, charge, and affinity for the stationary phase.
3. Ion Chromatography (IC): This technique separates ions in a solution by passing them through a column filled with an ion-exchange resin. The separation is based on differences in the ionic charge and affinity for the resin.
4. Size Exclusion Chromatography (SEC): This technique separates molecules based on their size and shape. A gel-filled column is used as the stationary phase, and the sample molecules are separated by their ability to pass through the gel pores.
5. Affinity Chromatography (AC): This technique separates molecules based on their specific binding to a ligand immobilized on a stationary phase. The ligand can be an enzyme, an antibody, or a receptor, and the separation is based on the strength and specificity of the binding interaction.

6. Thin Layer Chromatography (TLC): This technique separates compounds on a thin layer of stationary phase coated on a flat surface. The separation is based on differences in polarity, size, and affinity for the stationary phase.

Chromatographic techniques have been widely used in the separation, identification and detection of various compounds in different fields such as biological, food, pharmaceuticals, environmental samples (Kanu, 2021; Rodriguez-Mozaz et al., 2007; Malik et al., 2010).

There are several types of detectors that can be coupled to chromatography to detect and quantify the compounds. The widely used detectors can be listed as; UV-visible or diode array, electrochemical, refractive index, fluorescence, flame ionization and thermal conductivity and mass spectrometer (MS) detectors. Among these detectors, MS detector is the most widely used detector in chromatographic analysis due to its ability to provide information about the molecular weight and fragmentation patterns of compounds. Additionally, MS detectors can provide high sensitivity and selectivity for certain analytes, making them suitable for trace analysis and detection of compounds in complex matrices.

Preconcentration process is necessary for improving the sensitivity of the analytical detection method and improving the detection limits. The preconcentration factor (PF) is a quantitative measure that characterizes the extent of concentration or enrichment achieved for a specific analyte during the process of sample preparation. It provides insight into the degree of increase in analyte concentration relative to the initial sample concentration.

The PF is generally determined by comparing the volume (or mass) of the initial sample with the volume (or mass) of the final concentrated phase. A higher PF indicates a greater level of concentration or enrichment. The PF is mathematically expressed as the ratio of the initial sample volume (or mass) to the volume (or mass) of the final concentrated phase. It is commonly determined by dividing the volume or mass of the initial sample by the volume or mass of the final concentrated phase. PF provides a quantitative measure of the fold increase in analyte concentration achieved during the sample preparation process.

By employing various sample preparation techniques, the analyte can be selectively separated and concentrated from a complex matrix. The preconcentration step is crucial for enhancing the detection or analysis of the analyte, especially when it is present at low levels or when interference from other components is significant.

Sample preparation is an important step in chromatography that involves the extraction, separation and preconcentration of the analytes from the sample matrix to improve the sensitivity and selectivity of the analysis. There are various extraction techniques used in chromatography but the most widely used techniques are liquid-liquid extraction (LLE) and solid-phase extraction (SPE).

Liquid-Liquid Extraction (LLE)

Liquid-liquid extraction (LLE) is based on the analyte partitioning between water and an immiscible organic solvent. LLE is a chemical separation technique used to extract a target compound from a solution. This method is widely used in many fields, including analytical chemistry, biochemistry, pharmaceuticals, and the food industry. When used as a primary separation method, LLE involves the transfer of the target compound from one solution to another. The target compound or analyte selectively partitioned into one of the layers, allowing it to be separated from the other components. The choice of solvent and the extraction conditions can be optimized to achieve the maximum selectivity and efficiency. LLE can also be used as a purification step to isolate the target compound from impurities or to preconcentrate the analyte (Smith, 2003; Vaghela et al., 2016). While LLE is a cost-effective and efficient method of separation, it has two main limitations such as; it is more difficult to automate than SPE, which can limit its applicability in high-throughput settings and LLE may require the use of large volumes organic solvents. Despite these challenges, LLE remains a valuable tool for many chemical separation applications and can offer superior selectivity and efficiency (Dmitrienko et al., 2020; Masini et al., 2021). Reducing the amount of organic solvents used in sample preparation lowers the costs for their purchase and their disposal. Thus, various microextraction techniques have been developed. The use of alternative microextraction methods such as liquid-liquid microextraction (LLME) in sample preparation decreases the number of errors that usually occur in conventional multistage procedure. LLME requires smaller volumes of organic solvent, reduces extraction time, and minimizes the amount of emulsion formation when compared to LLE (Dmitrienko et al., 2021).

Hollow-fiber microextraction (HFME) method involves the use of a hollow fiber membrane as the extraction phase, which is typically coated with a selective sorbent material. The fiber is then placed in contact with the sample matrix, and the target analyte selectively partitioned into the membrane phase, while interfering compounds are excluded. HFME is a simple and efficient technique that offers several advantages over traditional sample preparation methods, including the ability to extract target analytes from small sample volumes, the potential for automation, and the ability to perform extractions in situ, without the need for sample transfer. Moreover, the high surface area-to-volume ratio of the hollow fiber membrane allows for the extraction of trace-level analytes (Bayatloo et al., 2022; Rasmussen and Bjergaard, 2004).

Solid-Phase Extraction (SPE)

Solid-phase extraction (SPE) is the most widely used sample preparation technique for the separation and preconcentration of tracel levels of compounds. SPE involves the use of a solid-phase sorbent material, usually packed into a column or cartridge, to selectively extract and concentrate target analytes from a complex sample matrix. Generally, the sample is typically passed through the SPE column, allowing the target analytes to selectively interact with the sorbent material. Meanwhile, interfering compounds are washed away, resulting in the separation of the target analytes from the sample matrix. SPE offers several advantages over traditional LLE techniques, including better reproducibility, improved selectivity and sensitivity, and reduced solvent consumption. Especially, over the last decade, various modifications have been made in the sorbents used in SPE to enhance the separation and enrichment performance of SPE. The main drawbacks of SPE are the time-consuming procedures required for isolating the analytes and the use of potentially toxic solvents. To overcome these problems, solid-phase microextraction (SPME) have been developed to separate and enrich target analytes from a sample matrix without the need for large volumes of solvents or additional steps in the sample preparation process (Buszewski and Szultka, 2021; Belmonte et al., 2022; Verma et al., 2023).

As a conclusion, the sample preparation step is very crucial in the analytical methods in order to improve the sensitivity, selectivity, accuracy and precision. Especially LLME techniques based on the use minimum amount of green or less toxic solvents, ionic liquids and deep eutetic solvents and novel or newly synthesized organic, inorganic and composites have been used as adsorbents or in the SPE systems for the preconcentration and determination of different types of organic substances prior to chromatographic analysis. This chapter will focus on the development of preconcentration techniques coupled with chromatographic methods for separation and determination of organic compounds.

Applications

Number of the chromatographic methods including enrichment or preconcentration step prior to analysis is increasing day by day. Pesticides, pharmaceuticals, volatile/aromatic compounds and their metabolites are the prominent organic compounds analyzed by the chromatographic instruments. Various types of techniques based on the usage of different solvents, organic/inorganic adsorbents or composite materials have been used in the preconcentration step of the analysis. In this perspective, some applications for the usage of preconcentration techniques prior to chromatographic analysis of organic compounds were given below.

Upon the various and extensive use of pharmaceuticals, many methods have developed for the analysis of the residuals in the real samples. In order to detect the trace levels, novel preconcentration techniques were incorporated to chromatographic methods.

Aghaei et al., (2019) fabricated an electromembrane extraction system to preconcentrate ampicillin in cow milk prior to HPLC-UV detection. High preconcentration factor (PF) (295) was obtained using octan-1-ol, silver nanoparticles, and reduced graphene oxide as a supported liquid membrane.

Ghoraba et al., (2017) synthesized zinc(II)-modified sulfur nanoparticles for the separation and preconcentration of carbamazepine. The research group used HPLC-UV to detect the analyte in pharmaceutical tablets, human plasma and breast milk. The researchers proposed that the nanoparticles were reusable and the developed method was simple, low-cost and consumed small amounts of organic solvent.

In the perspective of green chemistry, the use of less toxic solvents in the sample preparation or preconcentration step is crucial.

Ezoddin et al., (2019) fabricated a green method for the preconcentration of methadone based on the supramolecular solvent prior to gas chromatography–mass spectrometry (GC–MS). Supramolecular solvent system consists of the reverse micelles of 1-dodecanol in tetrahydrofuran and ultrasonic assisted supramolecular based on the solidification of floating organic drop micro extraction was used to detect the drug in human plasma and saliva samples.

Ionic liquids have been also used as a green solvent for the analysis of pharmaceuticals. Two different magnetic ionic liquids were synthesized by Hassan et al., (2021) to analyse carbamazepine drug in urine and environmental water samples. (Z)-octadec-9-en-1-aminium tetrachloroferrate (III) and (Z)-octadec-9-en-1-aminium trichlorocobaltate (II) were synthesized and utilized as sorbents for dispersive micro-solid phase extraction of the drug. Researchers proposed that the developed method offered simple, sensitive, efficient, and selective determination of the analyte.

Taghizadeh et al., (2022) also used magnetic ionic liquids for the preconcentration of five antidepressants. The research group synthesized six different magnetic ionic liquids consisting of different metal ions were and then hollow fiber lumen was injected as acceptor phase to extract the antidepressants. Magnetic ionic liquids were used both as the stirrer bar and as the extraction phase. Preconcentration factors higher than 294.3 and the method was applied to human milk and urine samples.

With the advantage of the mass spectrometry (MS) systems, multi-analysis can be achieved with low detection limits. Combining the MS/MS systems with the suitable preconcentration technique, multi-residue analysis with improved detection limits can be achieved. Idder et al., (2013) proposed a solid phase extraction technique based on the column system and residues of 40 drugs were investigated in various water samples.

Krakkó et al., (2019) analysed ten pharmaceuticals in water samples using Q-TOF/MS after preconcentration with the solid phase separation system. The research group combined the high-resolution of the mass spectrometric method with the high preconcentration factor (1250) obtained from the enrichment step.

Archana et al., (2016) combined solid-phase extraction combined with reversed-phase HPLC system with photodiode array detector for the analysis of different compounds including pharmaceuticals. Solid-phase extraction was carried out using a C18 cartridge.

Especially in recent years, deep eutectic solvents (DESs) have been widely used in the preconcentration step in order to develop green and environmental friendly methods.

According to literature, deep eutectic solvent can be defined as the mixture of two or more components at a certain molar ratio which gives the lower melting point compared to the components (Sekharan et al., 2022, Smith et al., 2014; Paiva et al., 2014). Deep eutectic solvents can be synthesized via the complexation of a hydrogen bond acceptor (HBA) such as quaternary ammonium salts and a hydrogen bond donor (HBD) such as alcohol, acid or carbohydrate (El Achkar et al., 2021; El-Deen and Shimizu, 2019). DESs generally have typical properties of being economical, biodegradable and non-toxic (Hansen et al., 2021).

Liu et al., (2019) proposed a method based on the usage of ultrasound-assisted dispersive liquid-liquid microextraction method based on a hydrophobic deep eutectic solvent for preconcentration of pyrethroid insecticides in water samples prior to HPLC-UV detection. DES was synthesized using quaternary phosphonium salts and straight-chain monobasic acids. Enrichment factors were obtained in the range of 127-131 for five insectides.

Kachangoon et al., (2020) synthesized DES using choline chloride as hydrogen bond acceptor and phenol as hydrogen bond donor in order to preconcentrate trace levels of four different neonicotinoid insecticides in water, soil and urine samples. DES was used as a disperser solvent in the extraction procedure and research group analysed the pesticides using HPLC-DAD and achieved high enrichment factors in the range of 50-250.

A deep eutectic solvent-based dispersive liquid–liquid microextraction method was combined with HPLC-MS for the preconcentration, separation and detection of pesticides by Wang et al., (2022). Six types of DES were synthesized using the same hydrogen bond acceptor thymol but with different hydrogen bond donors; decanoic, hexanoic, butyric undecylenic acids. With the aid of the preconcentration step, high enrichment factors were obtained in the range of 19.3-98.1.

Four different antibiotics were preconcentrated with the magnetic separation of chromium dioxide particles with a fatty acid surface layer which was applied to hydrophobic DES phase collection after liquid-liquid microextraction technique by Pochivalov et al., (2023). Thymol was mixed with different fatty acids in order to obtain DES. The method was applied to river and lake water samples to analyse trace levels of antibiotics by high-performance liquid chromatography with fluorometric detection. Upon the selection of the appropriate DES in the preconcentration step, high enrichment factors (95-97) were provided.

Considering their widespread use and toxicity, pesticides are one of the most analyzed organic substances. As trace levels of may remain in the environment for extended periods of time, development of novel and new preconcentration methods is of utmost importance.

With the combination of chromatographic methods and preconcentration techniques, it is possible to analyze not only a single pesticide but also a large number of pesticides at the same time.

Buarque et al., (2021) generated aqueous two-phase systems consist of tetrahydrofuran and glycerol for the preconcentration of atrazine. Aqueous two-phase systems can be prepared using two compounds when dissolved above a given concentration in aqueous media forming two immiscible aqueous liquid phases. Research group suggested that atrazine migrates to tatrahydrofuran-rich phase through hydrophobic interactions. LC-TOF-MS system was used to analyse atrazine in water samples and the preconcentration factor of 250 was obtained in the study.

Buppasang et al., (2022) developed a method based on the formation of a double solvent supramolecular system for enrichment of five triazole pesticides in milk samples. High enrichment factors of 73–318 was obtained using 1:1 of 1-undecanol/1-dodecanol which achieved the best extraction efficiency prior to HPLC-DAD detection.

Vichapong et al., (2022) fabricated a ringer tablet based micelle mediated extraction solvent back extraction method which is based on the entrapment of the analytes in the micelles of non-ionic surfactant, Triton X-114. Triton X-114 was used to extract pesticides into the surfactant rich phase, and then the analytes were back-extracted with 1-octanol. Four pesticides were determined in honey, cucumber and orange samples with HPLC-DAD.

Vortex assisted surfactant enhanced emulsification microextraction has been developed by Santaladchaiyakit et al., (2019) for the preconcentration and determination of benzimidazole fungicides in fruit juice samples. Methyl benzoate was used as extraction solvent and sodium dodecyl sulfate was used as emulsifier in the microextraction step. Pesticides were determined with a HPLC-DAD system.

GC-FID detection combined with a dispersive liquid-liquid microextraction method using a mixture of iso-propanol as a disperser and 1,2-dibromoethane as an extraction solvent for preconcentration of several widely used pesticides in fruit juices (Farajzadeh et al., 2021).

Use of the adsorbent plays an important role in the preconcentration of organic compounds. Preparation and modification of the novel adsorbents is a crucial step for the development of a sensitive and selective method.

Magnetic nanoporous carbon has been used for the adsorption and preconcentration of organophosphorus pesticides in different fruit samples (Li et al., 2019). The adsorbent was synthesized through direct carbonization of ZnO, $Co(OH)_2$ and 2-methyl imidazole mixture under inert atmosphere. Acetone/ethyl acetate (1:1 v/v) was used for the desorption of the pesticides, GC-FPD was used for the analysis and enrichment factors were obtained in the range of 56.2-77.9.

o-hydroxyazobenzene porous organic polymer also was used as an adsorbent for the adsorption of hormones and neonicotinoid insecticides (Selahle et al., 2022). The adsorbent was prepared via diazo coupling reaction. Upon the solid phase extraction, trace levels of analytes were analyzed in river water with HPLC-DAD.

Ilktac and Gumus (2022) synthesized magnetite-molecularly imprinted polymer using methacrylic acid and ethylene glycol dimethacrylate for the determination of pirimicarb. The pesticide was determined with the highly sensitive LC/Q-TOF/MS. The proposed method was applied to tap water and apple sample with a preconcentration factor of 5.

Santaladchaiyakit et al., (2021) fabricated the layered-double hydroxides with sodium dodecylbenzene sulfonate surfactant and preconcentrated and detected six benzoylurea insecticides in soft drink samples using HPLC-PDA. With the aid of anionic surfactant, hydrophilic surface was converted to a hydrophobic surface which enhanced the adsorption efficiency. The surfactant phase was dissolved in concentrated formic acid and high enrichment factors (38– 69) were obtained in the analysis.

Volatile organic compounds (VOCs) are the most common classes analyzed by chromatographic methods. However, these compounds are typically present in low concentrations in environmental samples, and preconcentration techniques enable their detection at trace levels. In order to improve the selectivity, sensitivity and detect the trace amounts of compounds, preconcentration techniques were developed prior to detection procedure. Thus, preconcentration is an important step in chromatographic analysis of VOCs. Generally, methods based on the GC systems were used for the determination of VOCs and solid-phase microextraction (SPME) and thermal desorption (TD) are two commonly used preconcentration techniques in VOCs analysis.

Bang et al., (2019) prepared a carbon nanotube sponge for the adsorption of VOCs including benzene, toluene, ethylbenzene, and o-xylene. Rapid thermal desorption was conducted after sorption and the VOCs were determined using GC-FID with the enrichment factors in the range of 200-300.

Polydimethylsiloxane polymer coated filter paper was used to preconcentrate benzene, toluene, ethylbenzene, and m-xylene before GC-MS analysis (Kim and Lim, 2019). With the proposed adsorbent and the headspace gas sampling method, sub-ppb levels could be detected in water samples.

Rodinkov et al., (2019) investigated the analytical use of commercial carbon based adsorbents for preconcentration of various VOCs from wet gas atmospheres prior to GC-FID detections. Carbon fluoroplastic composite sorbent was selected and the analytes were desorbed via thermal route and makes it possible to detect alcohols and ketones in air.

Rodinkov et al., (2021) also stated that various types of carbonaceous sorbents could be used for preconcentration VOCs. The research group prepared different composite adsorbents including silica supports such as; silica, aerosilogel, and diatomite and carbon materials such as; multiwall carbon nanotubes, graphitized thermal soot and pyrolytic carbon. With the adsorbent composed of diatomite porochrome-3 support coated with a pyrocarbon layer, phenol and isomeric cresols were detected upon their thermal desorption with GC-FID with the preconcentration factor of 2000.

Sharifkhani et al., (2022) developed a new technique for detecting the presence of certain organic compounds (benzene, toluene, ethylbenzene, and xylene) using an organic gas steam-liquid extraction method. After preconcentration step, analytes were detected using GC-FID. The preconcentration factors for these compounds were found to range from 118 to 207.

Various number of volatile compounds which can be considered as polycyclic aromatic hydrocarbons (PAHs) have been determined using GC or LC based methods. Preconcentration is an essential step in chromatographic analysis of PAHs. These compounds are typically present in low concentrations in environmental samples, and preconcentration techniques enable their detection at trace levels. Solid-phase microextraction (SPME) and liquid-liquid extraction (LLE) are two common preconcentration techniques used in PAH analysis. Overall, preconcentration is a crucial step in PAH analysis, as it enables the detection and quantification of these compounds at low concentrations in complex environmental matrices. With the aid of the MS systems, generally a group of analytes were analyzed at the same time.

Ma et al., (2020) fabricated mesoporous covalent organic polymer nanospheres for the preconcentration of PAHs prior to GC-MS analysis. The proposed method was applied to environmental water samples to detect naphthalene, fluorene, an- thracene, 1-nitronaphthalene, 2-nitrofluorene, 9-nitroanthracene, 9-fluorenone, and anthraquinone. The adsorbent was synthesized through a simple AlCl3-catalyzed Friedel-Crafts reaction between p-quaterphenyl and trimesoyl chloride and used in SPME. High PFs were obtained as 248–799 and limit of detection values in ng L-1 levels.

In the perspective of green chemistry, Rezaee et al., (2010) proposed a method that combined supercritical fluid extraction with DLLME for determination of ten different PAHs in marine sediment samples with the high

PF of 88-286. Chlorobenzene and acetonitrile were selected as the extraction and disperser solvent, respectively. The analytes were detected with GC–FID.

A novel method was designed by Li et al., (2008) for preconcentration and determination of phenanthrene, pyrene, chrysene, benzo[k]fluoranthene and benzo[a]pyrene in soil samples. Cloud point extraction was used as a preconcentration step and nonionic surfactant Tergitol 15-S-7 was used as the surfactant-mediated extractant. After preconcentration, HPLC equipped with a fluorescence detector was used for the analysis of the PAHs.

Polymeric monolith synthesized through a thermal polymerization reaction by using a mixture of n-butyl methacrylate as non-polar monomer and ethylene glycol dimethacrylate as crosslinker was used as an adsorbent for analysis of 15 different PAHs (Critto et al., 2022). The optimal conditions were determined as; n-butyl methacrylate as monomer, tetrahydrofurane (THF) as solvent for sorbent cleaning and THF: acetone (1:1) as elution solvent. PFs from 76 to 103 folds were obtained and the method was applied to river water.

The detection of other organic compounds and their metabolites, in addition to mentioned major classes, was achieved through the use of preconcentration strategies prior to chromatographic methods.

A group of catecholamine hormones (adrenaline, nor-adrenaline, and dopamin) were preconcentrated using hypercrosslinked polystyrene by Tolmacheva et al., (2019). The polymer was used in a microcolumn and the analytes were eluted with 1 mL of 6 M acetic acid and determined by reversed-phase HPLC with an amperometric detector with the PF of 20.

Investigation of preconcentration of adrenaline, nor-adrenaline, dopamine and their various metabolites in human urine samples using different extraction protocols were executed by Miękus et al., (2020). All separation studies were carried out using a capillary electrophoresis coupled with a DAD system. The researchers proposed that the developed method supported by SPE or SPME, can be used as an off-line preconcentration technique for the determination of catechol compounds in urine samples.

Activated carbon/octadecyl group modified silica monolithic discs were used for the preconcentration of esters in wine prior to GC-FID detection by Vyviurska et al., (2019) using hexane was used as the eluent.

Yu et al., (2022) proposed a novel two-phase amphiphilic preconcentration step based on the use of surfactants for preconcentration and analysis of phenolic compounds. Kaempferol, quercetin, salvianolic acid C and salvianolic acid B were analyzed using micellar electrokinetic chromatography. Cetyltrimethylammonium chloride (CTAC) and sodium dodecyl sulfate (SDS) surfactants were used in the preconcentration step. In the sample solution, the cationic surfactant CTAC was mixed with the anionic analytes before injection. Upon injection of the sample solution into the capillary, a boundary was created between CTAC and SDS in the background solution, causing the analytes bound inside micelles to be released due to the stronger electrostatic force between SDS and CTAC. Preconcentration step supplied PF in the range of 39–93.

Esrafili et al., (2012) fabricated an on-line hollow fiber liquid-phase microextraction method coupled to HPLC-UV for the preconcentration and analysis of pyridinic compounds; pyridine, 3-Methyl pyridine, 2,4-Dimethyl pyridine and quinolin in cigarette smoke sample with the PF in the range of 40-220. Dihexyl ether was selected as solvent for the extraction and a commercial hollow fiber was used in the study.

Seven polyphenols (chlorogenic acid, esculetin, caffeic acid, scopoletin, rutin, quercetin hydrate and kaempferol) was preconcentrated with a styrene-divinylbenzene resin (XAD-4) prior to reversed phase HPLC-DAD (Liu et al., 2008). The appropriate pH for adsorption was 2.5 and the analytes were eluted using methanol.

A preconcentration method based on the usage of phosphatidylcholine supramolecular solvent dispersive liquid-phase microextraction were designed by Tie et al., (2021) for the analysis of flavonoids; myricetin, quercetin, isorhamnetin, baicalein, wogonin and kaempferide in herbal aqueous samples. In the study, a supramolecular solvent formed by self-assembly of an amphoteric biosurfactant phosphatidylcholine and chloroform was used as the extraction phase and PF in the range of 107-205 was obtained.

Conclusion

This chapter focuses on the preconcentration strategies coupled with chromatographic methods for determination of various organic compounds. Preconcentration step not only minimizes the interference or matrix effects but also improves the sensitivity and detection limit of the analytical method. By combining the preconcentration step with the separation power of chromatography coupled with sensitive detectors, simultaneous analysis of trace levels of multiple compounds can be performed. Thus, the number of the chromatographic methods including preconcentration step is increasing day by day. However, the significant issue in preconcentration techniques is the high consumption of organic or toxic solvents. Therefore, future studies are likely to concentrate on the principles of green chemistry, aiming to address this concern and promote the usage of environmentally friendly adsorbents and practices. Additionally, in the pursuit of developing cost-effective systems, it is anticipated that there will be an increased focus on synthesizing more reusable adsorbents which enhance the efficiency and sustainability of the adsorption processes, ultimately reducing the overall cost associated with the use of adsorbents.

References

Aghaei A, Erfani Jazi M, E Mlsna T, Kamyabi MA. A novel method for the preconcentration and determination of ampicillin using electromembrane microextraction followed by high-performance liquid chromatography. *J. Sep. Sci.* 2019;42(18):3002-3008. doi:10.1002/jssc.201900016.

Archana G, Dhodapkar R, Kumar A. Offline solid-phase extraction for preconcentration of pharmaceuticals and personal care products in environmental water and their simultaneous determination using the reversed phase high-performance liquid chromatography method. *Environ. Monit. Assess.* 2016;188(9). doi:10.1007/s10661-016-5510-1.

Bang J, You DW, Jang Y, Oh JS, Jung KW. A carbon nanotube sponge as an adsorbent for vapor preconcentration of aromatic volatile organic compounds. *J. Chromatogr. A.* 2019;1605:460363. doi:10.1016/j.chroma.2019.460363.

Bayatloo MR, Tabani H, Nojavan S, Alexovič M, Ozkan SA. Liquid-Phase Microextraction Approaches for Preconcentration and Analysis of Chiral Compounds: A Review on Current Advances. *Crit. Rev. Anal. Chem.* 2022;1-15. doi:10.1080/10408347.2022.2038072.

Belmonte I dos S, Pizzolato TM, Gama MR. Quaternary ammonium pesticides: A review of chromatography and non-chromatography methods for determination of pesticide residues in water samples. *Trends Environ. Anal. Chem.* 2022;35. doi:10.1016/j.teac.2022.e00171.

Buarque FS, Lima NS, Soares CMF, Marques MN, Cavalcanti EB, Pereira MM, Souza RL, Lima AS. Preconcentration and chromatographic detection of atrazine in real water sample using aqueous two-phase system based on tetrahydrofuran and glycerol. *Environ. Qual. Manag.* 2021;31(1):39-48. doi:10. 1002/tqem.21738.

Buppasang R, Palasak J, Kachangoon R, Ponhong K, Teshima N, Burakham R, Srijaranai S, Vichapong J. Simultaneous Preconcentration of Triazole Fungicide Residues Using In-Situ Coacervative Extraction Based on a Double-Solvent Supramolecular System Prior to High Performance Liquid Chromatographic Analysis. *Molecules.* 2022;27(19). doi:10.3390/molecules 27196273.

Buszewski B, Szultka M. Past, Present, and Future of Solid Phase Extraction: A Review. *Crit. Rev. Anal. Chem.* 2012;42(3):198-213. doi:10.1080/073739372011. 645413.

Coskun O. Separation Tecniques: Chromatography. *North Clin. Istanbul.* 2016;3(2):156-160. doi:10.14744/nci.2016.32757.

Critto EF, Medina G, Reta M, Acquaviva A. Determination of polycyclic aromatic hydrocarbons in surface waters by high performance liquid chromatography previous to preconcentration through solid-phase extraction by using polymeric monoliths. *J. Chromatogr. A*. 2022;1679:463397. doi:10.1016/j.chroma.2022. 463397.

Dmitrienko SG, Apyari VV., Tolmacheva VV., Gorbunova MV. Dispersive Liquid–Liquid Microextraction of Organic Compounds: An Overview of Reviews. *J. Anal. Chem.* 2020;75(10):1237-1251. doi:10.1134/S1061934820100056.

Dmitrienko SG, Apyari VV., Tolmacheva VV., Gorbunova MV. Liquid–Liquid Extraction of Organic Compounds into a Single Drop of the Extractant: Overview of Reviews. *J. Anal. Chem.* 2021;76(8):907-919. doi:10.1134/S1061934821080049.

El Achkar T, Greige-Gerges H, Fourmentin S. Basics and properties of deep eutectic solvents: a review. *Environ. Chem. Lett.* 2021;19(4):3397-3408. doi:10.1007/ s10311-021-01225-8.

El-Deen AK, Shimizu K. Deep eutectic solvent as a novel disperser in dispersive liquid-liquid microextraction based on solidification of floating organic droplet (DLLME-SFOD) for preconcentration of steroids in water samples: Assessment of the method deleterious impact on the environment using Analytical Eco-Scale and Green Analytical Procedure Index. *Microchem. J.* 2019;149(June):103988. doi:10.1016/ j.microc.2019.103988.

Esrafili A, Yamini Y, Ghambarian M, Ebrahimpour B. Automated preconcentration and analysis of organic compounds by on-line hollow fiber liquid-phase microextraction-high performance liquid chromatography. *J. Chromatogr. A*. 2012;1262:27-33. doi:10.1016/j.chroma.2012.09.003.

Ezoddin M, Adlnasab L, kaveh A afshari, Karimi MA. Ultrasonically formation of supramolecular based ultrasound energy assisted solidification of floating organic drop microextraction for preconcentration of methadone in human plasma and saliva samples prior to gas chromatography–mass spectrometry. *Ultrason Sonochem.* 2019;50(September 2018):182-187. doi:10.1016/j.ultsonch.2018.09.019.

Farajzadeh MA, Kiavar L, Pezhhanfar S. Development of a method based on dispersive liquid–liquid microextraction followed by partial vaporization of the extract for ultra–preconcentration of some pesticide residues in fruit juices. *J. Chromatogr. A*. 2021;1653:462427. doi:10.1016/j.chroma.2021.462427.

Ghoraba Z, Aibaghi B, Soleymanpour A. Application of cation-modified sulfur nanoparticles as an efficient sorbent for separation and preconcentration of carbamazepine in biological and pharmaceutical samples prior to its determination by high-performance liquid chromatography. *J. Chromatogr. B Anal. Technol. Biomed Life Sci*. 2017;1063(March):245-252. doi:10.1016/j.jchromb.2017.07.048.

Hansen BB, Spittle S, Chen B, Poe D, Zhang Y, Klein JM, Horton A, Adhikari L, Zelovich T, Doherty BW, Gurkan B, Maginn EJ, Ragauskas A, Dadmun M, Zawodzinksi TA, Baker GA, Tuckerman MA, Savinell RF, Sangoro JR. Deep Eutectic Solvents: A Review of Fundamentals and Applications. *Chem. Rev.* 2021;121(3):1232-1285. doi:10.1021/acs.chemrev.0c00385.

Hassan AA, Tanimu A, Alhooshani K. Iron and cobalt-containing magnetic ionic liquids for dispersive micro-solid phase extraction coupled with HPLC-DAD for the preconcentration and quantification of carbamazepine drug in urine and environmental water samples. *J. Mol. Liq.* 2021;336:116370. doi:10.1016/j.molliq. 2021.116370.

Idder S, Ley L, Mazellier P, Budzinski H. Quantitative on-line preconcentration-liquid chromatography coupled with tandem mass spectrometry method for the determination of pharmaceutical compounds in water. *Anal. Chim. Acta.* 2013;805:107-115. doi:10.1016/j.aca.2013.10.041.

Ilktaç R, Gümüş ZP. Magnetite-molecularly imprinted polymer based highly sensitive chromatographic method for preconcentration and determination of pirimicarb. *Int. J. Environ. Anal. Chem.* 2022;102(17):5847-5859. doi:10.1080/ 03067319.2020. 1804890.

Kachangoon R, Vichapong J, Santaladchaiyakit Y, Srijaranai S. Cloud-point extraction coupled to in-situ metathesis reaction of deep eutectic solvents for preconcentration and liquid chromatographic analysis of neonicotinoid insecticide residues in water, soil and urine samples. *Microchem. J.* 2020;152 (October 2019):104377. doi:10.1016/j.microc.2019.104377.

Kanu AB. Recent developments in sample preparation techniques combined with high-performance liquid chromatography: A critical review. *J. Chromatogr. A.* 2021;1654:462444. doi:10.1016/j.chroma.2021.462444.

Kim EA, Lim YY. Effective preconcentration of volatile organic compounds from aqueous solutions with polydimethylsiloxane-coated filter paper. *Microchem. J.* 2019;145(December 2018):979-987. doi:10.1016/j.microc.2018.12.010.

Krakkó D, Licul-Kucera V, Záray G, Mihucz VG. Single-run ultra-high performance liquid chromatography for quantitative determination of ultra-traces of ten popular active pharmaceutical ingredients by quadrupole time-of-flight mass spectrometry after offline preconcentration by solid phase extraction from drinking and river waters as well as treated wastewater. *Microchem. J.* 2019; 148(January):108-119. doi:10.1016/j.microc.2019.04.047.

Li CF, Wong JWC, Huie CW, Choi MMF. On-line flow injection-cloud point preconcentration of polycyclic aromatic hydrocarbons coupled with high-performance liquid chromatography. *J. Chromatogr. A.* 2008;1214(1-2):11-16. doi:10.1016/ j.chroma.2008.10.062.

Li D, He M, Chen B, Hu B. Metal organic frameworks-derived magnetic nanoporous carbon for preconcentration of organophosphorus pesticides from fruit samples followed by gas chromatography-flame photometric detection. *J. Chromatogr. A.* 2019;1583:19-27. doi:10.1016/j.chroma.2018.11.012.

Liu Q, Cai W, Shao X. Determination of seven polyphenols in water by high performance liquid chromatography combined with preconcentration. *Talanta.* 2008;77(2):679-683. doi:10.1016/j.talanta.2008.07.011.

Liu X, Liu C, Qian H, Qu Y, Zhang S, Lu R, Gao H, Zhou W. Ultrasound-assisted dispersive liquid-liquid microextraction based on a hydrophobic deep eutectic solvent for the preconcentration of pyrethroid insecticides prior to determination by high-performance liquid chromatography. *Microchem. J.* 2019;146(October 2018):614-621. doi:10.1016/j.microc.2019.01.048.

Ma R, Wang W, Wang Z, Zhang S, Li Z, Li J, Zang X, Wang X, Wang Z. Mesoporous covalent organic polymer nanospheres for the preconcentration of polycyclic aromatic hydrocarbons and their derivatives. *J. Chromatogr A*. 2020;1624:461217. doi:10.1016/j.chroma.2020.461217.

Malik AK, Blasco C, Picó Y. Liquid chromatography-mass spectrometry in food safety. *J. Chromatogr. A*. 2010;1217(25):4018-4040. doi:10.1016/j.chroma.2010. 03.015.

Masini JC, do Nascimento FH, Vitek R. Porous monolithic materials for extraction and preconcentration of pollutants from environmental waters. *Trends Environ. Anal. Chem*. 2021;29:e00112. doi:10.1016/j.teac.2020.e00112.

Miękus N, Plenis A, Rudnicka M, Kossakowska N, Oledzka I, Kowalski P, Baczek T. Extraction and preconcentration of compounds from the L-tyrosine metabolic pathway prior to their micellar electrokinetic chromatography separation. *J. Chromatogr. A*. 2020;1620. doi:10. 1016/j.chroma.2020.461032.

Paiva A, Craveiro R, Aroso I, Martins M, Reis RL, Duarte ARC. Natural deep eutectic solvents - Solvents for the 21st century. *ACS Sustain Chem Eng*. 2014;2(5):1063-1071. doi:10.1021/sc500096j.

Parker S. Principles and Practice. *IFLA J*. 2006;32(3):179-180. doi:10.1177/03400 35206070163.

Pochivalov A, Cherkashina K, Sudarkin A, Osmolowsky M, Osmolovskaya O, Krekhova F, Nugbienyo L, Bulatov A. Liquid-liquid microextraction with hydrophobic deep eutectic solvent followed by magnetic phase separation for preconcentration of antibiotics. *Talanta*. 2023;252 (August 2022):123868. doi:10.1016/j.talanta.2022. 123868.

Rasmussen KE, Pedersen-Bjergaard S. Developments in hollow fibre-based, liquid-phase microextraction. *TrAC - Trends Anal Chem*. 2004;23(1):1-10. doi:10.1016/ S0165-9936(04)00105-0.

Rezaee M, Yamini Y, Moradi M, Saleh A, Faraji M, Naeeni MH. Supercritical fluid extraction combined with dispersive liquid-liquid microextraction as a sensitive and efficient sample preparation method for determination of organic compounds in solid samples. *J. Supercrit. Fluids*. 2010;55(1):161-168. doi:10.1016/ j.supflu.2010.07.008.

Rodinkov OV., Vagner EA, Bugaichenko AS, Moskvin LN. Comparison of the Efficiencies of Carbon Sorbents for the Preconcentration of Highly Volatile Organic Substances from Wet Gas Atmospheres for the Subsequent Gas-Chromatographic Determination. *J. Anal. Chem*. 2019;74(9):877-882. doi:10.1134/S106193481909 0089.

Rodinkov O, Postnov V, Spivakovskyi V, Vlasov A, Bugaichenko A, Slastina S, Znamenskaya E, Shilov R, Lanin S, Nesterenko P. Comparison of adsorbents containing carbon nanotubes for express pre-concentration of volatile organic compounds from the air flow. *Separations*. 2021;8(4). doi:10.3390/separations 8040050.

Rodriguez-Mozaz S, Lopez de Alda MJ, Barceló D. Advantages and limitations of on-line solid phase extraction coupled to liquid chromatography-mass spectrometry technologies versus biosensors for monitoring of emerging contaminants in water. *J. Chromatogr. A*. 2007;1152(1-2):97-115. doi:10.1016/j.chroma.2007.01.046.

Santaladchaiyakit Y, Sadchaiyaphum J, Phannorit N, Lekyoung C, Chumwangwapee N, Srijaranai S. Modified Vortex-assisted Surfactant-enhanced Emulsification

Microextraction using Methyl Benzoate for the Preconcentration of Fungicides in Fruit Juice Samples and Determination by High-performance Liquid Chromatography. *J. Anal. Chem.* 2019;74(10):1027-1034. doi:10.1134/S106193481 9100095.

Santaladchaiyakit Y, Sila-Am W, Sribunrueng S, Gissawong N, Srijaranai S. Co-precipitation based on layered double hydroxides and anionic surfactants for preconcentration of six benzoylurea insecticides in soft drinks before simultaneous analysis by high-performance liquid chromatography. *Anal. Methods*. 2021;13(47): 5735-5748. doi:10.1039/d1ay01435h.

Sekharan TR, Chandira RM, Tamilvanan S, Rajesh SC, Venkateswarlu BS. Deep eutectic solvents as an alternate to other harmful solvents. *Biointerface Res. Appl. Chem.* 2022;12(1):847-860. doi:10.33263/BRIAC121.847860.

Selahle SK, Mpupa A, Nqombolo A, Nomngongo PN. A nanostructured o-hydroxyazobenzene porous organic polymer as an effective sorbent for the extraction and preconcentration of selected hormones and insecticides in river water. *Microchem. J.* 2022;181(May):107791. doi:10.1016/j.microc.2022.107791.

Sharifkhani SM, Yaftian MR, Hosseini MH, Zolfonoun E. Introducing Organic Gas Steam-Liquid Extraction as a New Preconcentration Method for Benzene, Toluene, Ethylbenzene and Xylene Determination in Water Samples by Gas Chromatography-Flame Ionization Detection. *J. Anal. Chem.* 2022;77(4):505-512. doi:10.1134/ S1061934822040098.

Smith EL, Abbott AP, Ryder KS. Deep Eutectic Solvents (DESs) and Their Applications. *Chem. Rev*. 2014;114(21):11060-11082. doi:10.1021/cr300162p.

Smith RM. Before the injection - Modern methods of sample preparation for separation techniques. *J. Chromatogr. A*. 2003;1000(1-2):3-27. doi:10.1016/S0021-9673(03) 00511-9.

Taghizadeh M, Ebrahimi M, Fooladi E, Yoosefian M. Preconcentration and determination of five antidepressants from human milk and urine samples by stir bar filled magnetic ionic liquids using liquid–liquid–liquid microextraction-high-performance liquid chromatography. *J. Sep. Sci*. 2022;45(8):1434-1444. doi:10. 1002/jssc.202100617.

Tie D, Chen X, Wang R qin, Hu S, Yang L, Zheng Y. Simultaneous preconcentration and determination of trace flavonoids in complex matrix by phosphatidylcholine supramolecular solvent-based dispersive liquid-phase microextraction. *Microchem. J.* 2021;168(April):106348. doi:10.1016/j.microc.2021.106348.

Tolmacheva VV., Yarykin DI, Gorbunova MV., Apyari VV., Dmitrienko S G, Zolotov YA. Preconcentration of Catecholamins on Hypercrosslinked Polystyrene and Their Determination by High-Performance Liquid Chromatography. *J. Anal. Chem.* 2019;74(11):1057-1063. doi:10.1134/S1061934819090107.

Vaghela A, Patel A, Patel A, Vyas A, Patel N. Sample Preparation In Bioanalysis : A Review. *Int. J. Sci. Technol. Res*. 2016;5(05):6-10.

Verma R, Dhingra G, Malik AK. A Comprehensive Review on Metal Organic Framework Based Preconcentration Strategies for Chromatographic Analysis of Organic Pollutants. *Crit. Rev. Anal. Chem.* 2023;53(2):415-441. doi:10.1080/ 10408347. 2021.1964344.

Vichapong J, Kachangoon R, Burakham R, Santaladchaiyakit Y, Srijaranai S. Ringer Tablet-Based Micelle-Mediated Extraction-Solvent Back Extraction Coupled with High-Performance Liquid Chromatography for Preconcentration and Determination of Neonicotinoid Pesticides. *Food Anal. Methods*. 2022;15(4):970-980. doi:10.1007/s12161-021-02067-4.

Vyviurska O, Gorovenko R, Panáková V, Špánik I. Extensive Validation of Gas Chromatographic Method for Determination of Esters in Wines Using Monolithic Materials as Preconcentration Step. *Food Anal. Methods*. 2019;12(3):791-798. doi:10.1007/s12161-018-01409-z.

Wang L, Wang Y, Chen M, Qin Y, Zhou Y. Hydrophobic deep eutectic solvent based dispersive liquid–liquid microextraction for the preconcentration and HPLC analysis of five rice paddy herbicides in water samples. *Microchem. J.* 2022;181(May):107790. doi:10.1016/j.microc.2022.107790.

Yu YL, Shi MZ, Zhu SC, Cao J. The two-phase amphiphilic preconcentration based on surfactants to enrich phenolic compounds from diluted plant extracts and rat urine by micellar electrokinetic chromatography. *Electrophoresis*. 2022;43(16-17):1735-1745. doi:10.1002/elps.202200044.

Chapter 3

Bioanalytical Methods Based on Liquid Chromatography for the Therapeutic Drug Monitoring of Combined Therapies of Antibiotics in Pediatric Septic Major Burns

Silvia Regina Cavani Jorge Santos[1,*], PhD
Thais Vieira Camargo[1], MS, PhD
Leonard de Vinci Kanda Kupa[1], PhD
Ronaldo Morales Junior[1], PhD
Claudia Garcia Messiano[1], MS
Vanessa Kasubeck de Souza[1], MS
Frederico Ribeiro Pires[1], MD
Aline Sandre[2], BS
Gabriela Aparecida Ferreira[2], BS
Edvaldo Vieira de Campos[2]
João Manoel da Silva Junior[2], MD
Débora Cristina Sanches Pinto[2], MD
and David de Souza Gomez[2,†], MD

[1]Clinical Pharmacokinetics Center, University of Sao Paulo, Sao Paulo, Brazil
[2]Plastic Surgery and Burns Division of Clinics Hospital,
Surgery Department of Medical School, University of Sao Paulo, Sao Paulo, Brazil

* Corresponding Author's Email: pharther@usp.br.
† Corresponding Author's Email: davgomez@usp.br.

In: Chromatographic Methods and Research
Editor: Raven W. Obrien
ISBN: 979-8-89113-193-4

Abstract

Meropenem and vancomycin combined therapy is largely prescribed for septic patients from Intensive Care Units (ICU). Therapeutic drug monitoring (TDM) of these antibiotics in those patients permits to evaluate drug effectiveness against infections caused by Gram-negative and Gram-positive strains by applying pharmacokinetic-pharmacodynamics tools based on targets recommended. Subject of study was to evaluate drug effectiveness by developing of a bioanalytical method for meropenem and vancomycin serum measurements by liquid chromatography equipped with ultraviolet detector (LC-UV). It is well known that liquid chromatography technique is considered the gold standard for TDM of antimicrobial agents due to its sensitivity and selectivity. Then, it was developed and validated a bioanalytical method to quantify both antibiotics in serum samples in ICU septic patients done by LC-UV set at 307 nm for meropenem and vancomycin, set at 230 nm to investigate drug effectiveness in major septic pediatric burns. Validation study was based on international standards considering the parameters as linearity, lower limit of quantification, lower limit of detection, intra-day/inter days precisions, intra-day/inter days systematic error, stability of standards on the bench (4 hrs.), short serum samples stability on the bench (6 hrs.), post-process stability of serum extracts in the rack of autosampler up to 24 hrs., serum samples long stability (ultralow freezer storage), cycles freezing/de-freezing stability, and the robustness of chromatographic system. Pediatric major burns with preserved renal function receiving a combined therapy with meropenem by 3 hrs.-infusion, and vancomycin by 1 hr. intermittent infusion were included. Patients were stratified by age in two groups receiving the meropenem 40 mg/kg q8h for group 1 (G1) n = 22 (2-13 yrs.) and 20 mg/kg q8h for group 2 (G2) n = 20 (14-18 yrs.). Vancomycin empirical dose regimens were 10-15 mg/kg q6h for G1-patients and 10-15mg/kg q8h for G2-patients. Blood was sampling 48 hrs. after the therapy started for drug serum measurements at the 3rd hr. of started infusion, and one hour before the next infusion. Pharmacokinetic-pharmacodynamics approach-based on serum levels to attain therapeutic targets recommended for vancomycin (AUC/MIC > 400-600), and for meropenem (100%$f\Delta T$ > MIC) were applied to evaluate antimicrobial effectiveness. Clinical cure occurred for all patients with eradication of Gram-negatives pathogens up to MIC 4 mg/L by meropenem, and of Gram-positive strains up to MIC 2 mg/L by vancomycin according to Clinical Standards Laboratory Institute, data base in our hospital.

Keywords: bioanalytical simultaneous analyses - LC-UV, therapeutic drug monitoring, antimicrobial combined therapy, PK/PD tools based on serum levels, target attainment by antimicrobial effectiveness, vancomycin target AUC/MIC > 400, meropenem target 100%fDT > MIC, therapy guided by PK/PD approach and cultures

Introduction

Septic shock is potentially fatal organ dysfunction caused by a dysregulated host response to infection. Clinical outcome in most high-risk cases is the mortality of patients including pediatrics with nosocomial bacterial infections associated with viral infections as influenza and most recently SARS-CoV-2 (Rudd et al., 2020). In view of the growing challenge to the prescription of antimicrobials for adequate treatment and effective control of bacterial infectious conditions during COVID-19 pandemic, International Guidelines for Management of Sepsis and Septic Shock (2021), Global Sepsis Alliance (2020), and World Health Organization (2023) have reinforced that combating bacterial resistance, and the prevention of the development of multidrug-resistant strains (MDR) is urgent (Abdulla et al., 2020; Elligsen, Walker and Simor, 2011; Evans et al., 2021; Greenhalgh et al., 2007). It is well known that at the last twenty years, it was reported a significant increase in the minimum inhibitory concentration (MIC). In a monitoring program of antimicrobial therapy for several infections based on serum levels of beta-lactams in ICU patients, it was reported that almost 80% of patients did not reach the therapeutic target by vancomycin against susceptible strains of Gram-positive strains and Gram-negative bacteria by meropenem. Information reinforces that antimicrobial serum levels monitoring is essential to assess pharmacokinetics changes that occurs during the septic shock with impact on coverage of the prescribed antimicrobial agent, expressed through target attainment reached by pharmacokinetic-pharmacodynamics (PK/PD) approach. Therefore, new therapeutic strategies have been proposed for the most prescribed beta-lactam agents related to the dose regimen and the duration of infusion mainly for meropenem (Abdul-Aziz et al., 2016; Barlam et al., 2016; Carlier et al., 2015; Roberts et al., 2010). Some prospective controlled studies including therapeutic drug monitoring (TDM) were considered to compare drug efficacy based on serum levels of several antimicrobial agents usually prescribed in the therapy of septic shock. Considering that therapeutic meropenem serum monitoring after 3 hrs.-extended infusion is a strategy that ensures an adequate serum level against intermediate susceptible strains with coverage up to MIC 4 mg/L in the combat against bacteria of intermediate susceptibility mainly related to *K. pneumoniae* and *P. aeruginosa* in major septic burns with renal function preserved or augmented by receiving vasopressors. In addition, to allowing the evaluation of vancomycin effectiveness considering the dose regimen prescribed to septic ICU patients, drug serum levels are a laboratory strategy of great value in the individualization of therapy in a real time, guaranteeing the expected clinical outcome, and to combating the development of bacterial resistance, also reducing the duration of antimicrobial therapy, and consequently hospital costs (Abdulla et al., 2020; Barlam et al., 2016).

Impact of the introduction of real-time therapeutic drug monitoring on empirical doses of carbapenems in critically ill burn patients was reported to adequate empirical antibiotic dose selection for critically ill burn patients, once is difficult due to extreme variability in drug pharmacokinetics. Therapeutic drug monitoring (TDM) by liquid chromatography may aid antibiotic prescription and implementation of initial empirical antimicrobial dosage recommendations. In this study it was evaluated how gradual TDM introduction altered empirical dosages of meropenem and imipenem/cilastatin in the ICU Burned Center of hospital (Fournier et al., 2015). Other studies related to meropenem and several beta-lactam agents TDM were reported by applying LC-UV (Caro et al., 2020, Casals et al., 2014; Ferrone et al., 2017; Fournier et al., 2015; MacWhinney et al., 2010; Santos et al., 2011; Zou et al., 2019). Controversially, other techniques based on immunoassays with radioisotope or fluorescein markers, and microbiological technical procedures for beta-lactams serum measurements were reported (Crossley et al., 1980; Fridlund, Woksepp and Shona, 2016). On the other hand, concerning bioanalytical methods for vancomycin serum measurements by applying liquid chromatography-ultraviolet detection, a revision of chromatographic procedures was reported (Javorska et al., 2016). Then, it was selected few studies related to LC-UV for vancomycin TDM in human serum or plasma with linearity and sensitivity required for a routine in central laboratory of hospitals (Cao et al., 2014; Demotes-Mainard, Labat and Vinc, 1994; Furuta et al., 2000; Hagihara, Sutherland and Nicolau, 2013; Lopez et al., 2007).

Finally, it is noteworthy that, so far, serum levels of these agents are not routinely monitored in hospitals for these patients admitted to ICUs (Abdul-Aziz et al., 2016; Carlier et al., 2015). Clinical management for critically ill patients in intensive care has been guided by cultures, through the isolation of the agent, and susceptibility test of the pathogen to the antimicrobial., supported by ICU routine of serum biomarkers of the systemic inflammatory response syndrome (SRIS). Then, if culture results and coverage by PK/PD tools based on meropenem and vancomycin serum levels, it is possible to guide properly the antimicrobial therapy to reach soon the desired clinical outcome. If serum levels of prescribed antimicrobial agent are equal to or greater than those required to eradicate the susceptible pathogens, microbiological cure occurs, and the desired clinical outcome will be achieved by appropriately treating the septic shock and healing the patient; on the other hand, for serum levels below the recommended level, therapeutic failure will occur contributing to development of microbial resistance and bacterial emergency (Abdul-Aziz, 2015; Elligsen, Walker and Simor, 2011; Abdulla et al., 2020).

Then, selective techniques as liquid chromatography must be chosen as laboratory strategy in hospital of high complexity to facilitate therapeutic drug serum monitoring done in a real time for ICU critically ill septic patients with nosocomial infections caused by Gram-positive and Gram-negative pathogens to reach soon desired outcome, contributing to combat microbial resistance.

Rational of Study

Subject of study was to evaluate drug effectiveness by application of pharmacokinetic-pharmacodynamics (PK/PD) tools based on serum levels done by liquid chromatography-ultraviolet detector (LC-UV) using a bioanalytical method developed for meropenem and vancomycin serum measurements once liquid chromatography technique is considered the gold standard for TDM of antimicrobial agents due to its sensitivity and selectivity.

Bioanalytical Method

To each 200 μL of meropenem serum sample, 25 μL (50 μg/mL) of cefepime (internal standard, US Pharmacopeia) was added to an Eppendorf tube followed by 600 μL of acetonitrile and centrifuged 8,000 rpm (5°C) for 15 min. Supernatant was evaporated to dryness in a water bath (45°C) under a stream of nitrogen. A volume of 600 μL of 10% of acetonitrile in water was added to dissolve the residue, and 5 μL of serum extracts were injected into a Shimpack ODS column (150 mm x 6 mm ID, 5 micron). Elution of peaks was done by an isocratic system with a mobile phase of a mixture of acetate buffer 0.01 M and acetonitrile (9:1, v/v) at a flow rate of 0.8 mL/min. Peaks were monitored by UV detection at 307 nm for meropenem, and at 230 nm for vancomycin. Validation study was based on international standards considering the following parameters as linearity, lower limits of quantification and of detection, intra-day/inter days precisions, intra-day/inter days systematic error, stability of standards including the internal standard on the bench during 4 hrs., short samples stability on the bench (6 hrs.), post-process stability of serum extracts in the rack of autosampler rack (24 hrs.), samples long stability storage in an ultralow freezer, cycles freezing/de-freezing stability, and the robustness of chromatographic system.

The clinical protocol was a prospective, open-label study. Ethical approval register CAAE 07525118.3.0000.0068, Brazilian Platform was obtained by approval of the Ethical Committee of Hospital of Clinics, Medical School, University of Sao Paulo; no conflicts of interest to declare were obtained from all authors. The study was conducted from January 2018 to March 2020, and informed written consent was obtained from all legally designated patient representatives. Pediatric patients from the Intensive Care Burn Unit aging 2 yrs. up to 18 years old, presenting severe injury by fire or electricity, and a sepsis diagnosis as reported in the "American Burn Association consensus conference to define sepsis and infection in burns" in clinical evaluation and laboratorial data were eligible for inclusion (Greenhalgh et al., 2007). Patients with vancomycin or meropenem intolerance or renal impairment were excluded. The study was based on the recommended antimicrobial treatment to suspected or documented Gram-positive and Gram-negative nosocomial infections of hospital.

Clinical Trial: Study Design and Patient Eligibility, Combined Therapy, Blood Sampling, PK/PD Approach

Thus, vancomycin-meropenem combined therapy were prescribed at dose regimens recommended for pediatric patients with renal function preserved or augmented by vasopressors requirements, that occurs mainly at the earlier stage of septic shock. Septic major burn pediatric patients 42 (31M/11F) at the earlier stage of septic shock were included to investigate antimicrobial effectiveness of the combined therapy guided by cultures, and PK/PD tools based on drug serum levels. Patients were allocated by age in two groups: 2-13 yrs. (Group 1, n = 22) and 14-18 yrs. (Group 2, n = 20). Initial dose regimen of vancomycin was given by one hour pump infusion in Set 1to G1-patients, body weight normalized, 10-15 mg/kg q6h, or 10-15 mg/kg q8h to G2-patients. Vancomycin dose adjustment was done in Set 2, if required to achieve clinical outcome purposes based on coverage of isolates from cultures. In addition, meropenem was administered by 3 hrs.- pump infusion at the dose regimen recommended, body weight normalized, 40 mg/kg q8h for G1-patients and 20 mg/kg q8h for G2-patients, according to the institutional protocol. Complete medical histories, physical examinations were obtained for each enrolled patient; laboratory data included documented strains in blood, bone, pulmonary lavage, and urine cultures; susceptibility testing done to obtain the minimum inhibitory concentration for each antimicrobial agent against each pathogen isolated.

It was investigated vancomycin effectiveness at empirical dose regimen based on the area under the curve and the minimum inhibitory concentration ratio ($AUC^{ss}_{0\text{-}24}/MIC > 400\text{-}600$), pharmacokinetics-pharmacodynamics (PK/PD) target recommended for critically ill septic patients at the last consensus (Rybak et al., 2020). Then, empirical dose regimen was done in Set 1 of the protocol to guarantee the target achievement, and if required dose adjustment was done in Set 2 to achieve vancomycin effectiveness against Gram-positive isolated from cultures. TDM data for individualization of vancomycin therapy in Set 2 was compared with results obtained in Set 1. Blood was sampling in both sets at the 3rd hr. of infusion started, and one hour before the next infusion for serum levels done by liquid chromatography. One compartment open model was chosen to investigate pharmacokinetics, and noncompartmental data analysis was applied. Drug effectiveness was evaluated by pharmacokinetic-pharmacodynamic approach, guided by PK/PD target attainment and microbiology of Gram-positive isolates from cultures.

In addition, all septic pediatric major burns received also the recommended meropenem regimen done by 3hrs.-infusion to attain the recommended target ($100\%f\Delta T > MIC$) previously reported (Abdul-Aziz et al., 2016). New target considered was related to the percentage of time dose interval, that the minimum inhibitory concentration was attained in ICU septic patients by beta lactam agents largely prescribed. Blood was sampling at the 3rd and 7th hr. of the infusion started for serum levels done by liquid chromatography. One compartment open model was chosen to investigate pharmacokinetics, and noncompartmental data analysis was applied. Drug effectiveness was guided by PK/PD approach, $100\%f\Delta T > MIC$ target considered. based on serum levels, and by the microbiology of Gram-negative isolates from cultures.

Analytical Procedures for Drug Serum Measurements

Table 1. Bioanalytical method for drug serum measurements by liquid chromatography

LC UV-VIS SHIMADZU LC10A		
Bioanalytical Method (serum matrices)	LC UV-Vis 307 nm	LC UV 230 nm
Validation parameters	Meropenem	Vancomycin
Linearity (lower/upper limits)	0.5-200 mg/L	1.0-200 mg/L
Linear regression coefficient	0.9998	0.9996
Lower limit f quantification (LOOQ)	0.5 mg/L	1.0 mg/L
Limit of detection (LOD)	0.25 mg/L	0.50 mg/L
Relative recovery (mean, CI95%)	98(94-103)	97(95-105)
Intraday precision (CV%) n = 18	6.1(5.5-6.8)	6.6(5.9-7.2)
Interday precision (CV%) n = 27	7.8(6.5- 9.3)	7.1(6.0- 8.9)
Systematic error intra-day (%) n = 18	7.8(6.9-8.5)	5.8(4.9-7.6)
Systematic error interday (%) n = 27	8.8(7.9-9.8)	6.7(6.1-8.7)
Standard stability, bench 4 hrs. (%) n = 9	2.8(1.9-4.5)	3.4(2.6-5.2)
Internal standard stability, bench 4 hrs. (%) n = 6	3.5(3.1-4.5)	2.5(2.1-4.4)
Samples short stability, bench 6 hrs. (%) n = 9	4.5(2.8-4.2)	3.9(3.3-5.0)
Post-process stability, rack 24 hrs. (%) n = 9	5.8(4.2-6.5)	4.8(4.0-7.2)
Samples long stability, UL freezer (%) n = 118	7.5(7.2-9.4)	6.9(6.2-8.8)
Cycles freezing/de-freezing stability n = 27	5.8(5.5-6.7)	4.9(4.5-5.9)
Robustness (chromatographic system) n = 72	8.1(6.2-8.7)	7.2(6.9-8.9)

Abbreviations: CV: coefficient of variation; LOOQ: lower limit of quantification; LOD: Limit of detection; CI: Confidence interval; UL: ultralow.
Statistics: GraphPad Prism software, version 9.1.4.

Considering meropenem serum measurements, a carbapenem agent prescribed to combat infections caused by Gram-negative strains, confidence limits are described in Table 1 including a linear response of 0.5 up to 200 mg/L (r^2:0.9998) that was obtained with a lower limit of quantification (LLOQ) of 0.5 mg/L, and a limit of detection 0.25 mg/L. Intraday, and inter days precisions were acceptable at low, medium, and high concentrations based on internal controls, and the relative recovery ranging 94-103% was obtained. It was evidenced acceptable stability related to standards on the bench (4 hrs.), samples short stability (6 hrs.) on the bench, post-process stability 24 hrs. in the autosampler, samples long stability by storing in the ultralow freezer, and cycles freezing/de-freezing stability. Robustness of each chromatographic system was also included. Finally, LC-UV was successfully developed and validated with a run time lower than 10 min. per sample analyzed in a rack of 100 mini vials overnight.

Also, for vancomycin, a glycopeptide agent largely prescribed to treat infections caused by Gram positive bacteria, the confidence limits of the bioanalytical method validated were described also in Table 1. It was registered a linearity from 1.0 up to 200 mg/L (r^2:0.9996), lower limit of quantification (LLOQ) of 1.0 mg/L, and a limit of detection 0.5 mg/L were obtained. Intraday, and inter days precisions were acceptable at low, medium, and high concentrations, and the relative recovery ranged between 95-105%. It was evidenced accepted stability related to standards on the bench (4 hrs.), samples short stability (6 hrs.) on the bench, post-process stability for 24 hrs. in the autosampler, samples long stability in the ultralow freezer, and cycles freezing/de-freezing stability. Robustness of each chromatographic system was also included. Finally, LC-UV was successfully developed and validated with a run time lower than 10 min. per sample analyzed in a rack of 100 mini vials overnight. Vancomycin serum levels from patients were determined by LC-UV 230 and by immunoassay in the Cobas Integra 8,000 analyzer (Roche Diagnostics, Switzerland) with fluorescence polarization immunoassay in the central laboratory of hospital. Linearity range for vancomycin serum levels by the immunoassay was from 4 up to 80 mg/L, with lower limit of quantification of 4 mg/L (LLOQ)), and coefficient of variation of 4% over the entire calibration daily curve. Despite the reduced period for processing vancomycin serum levels by automatized immunoassay analyzer, Cobas Integra 8,000 of central laboratory of hospital, the superiority of LC-UV technique over immunoassay technique related to the confidence limits was evidenced due to a large linearity and high sensitivity based on LLOD of 1 mg/L of the liquid chromatographic method, that justifies the results obtained for this gold standard technique to quantify antimicrobial agents for TDM. Antimicrobials serum samples of patients were analyzed after procedure of precipitation of serum proteins with acetonitrile for vancomycin and meropenem. Purified extracts (5 μL, volume) were injected automatically overnight and analyzed by a LC-UV detection at 230 nm for vancomycin, and 307 nm for meropenem in a liquid chromatograph LC10A, Shimadzu Corporation (Kyoto, Japan). Only 0.2 mL of each sample was required for drug serum measurements, replicates $n = 2$, using high performance liquid chromatography/ultraviolet detection (LC-UV) according to the simple and accurate bioanalytical methods with adequate linearity, sensitivity, accuracy, and precision; stability studies were also considered. Coefficient of determination ($r2 > 0.999$) for the drug assay over the standard curve concentrations based on eight serum calibrators plus the zero (C0) were acceptable (linearity C1-C8: up to 200 mg/L), C1:1 mg/L (vancomycin) or C1: 0.5 mg/L (meropenem). Additionally internal controls, high, medium, and low were included. Calibration daily curve was accepted based on systematic error lower than 10% for the internal controls (high, medium, and low serum levels); then, drug serum levels in patients' samples were determined based on the accepted daily curve using the internal standard method, Table 1.

Clinical Study

Table 2. Demographic characteristics of pediatric septic burn patients admission in ICU, biomarkers at TDM combined therapy, medians (quartiles)

Demographic data	G1 n1 = 22 (2-13 yrs.)	G2 n2 = 20 (14-18 yrs.)	p
Gender (Male/Female) proportion	15M/7F	16M/4F	$p > 0.05$
Age (yrs)	5.0(3.2-9.8)	16.0(14.4-17.6)	$p < 0.05$
Body weight (kg)	22(16-38)	65(55-66)	$p < 0.05$
Heights (cm)	118(108-134)	165(160-166)	$p < 0.05$
ICU admission data			
TBSA %	30(20-39)	36(27-48)	$p > 0.05$
Inhalation injury (proportion)	16/22	15/20	$p > 0.05$
Mechanical ventilation (proportion)	16/22	18/20	$p > 0.05$
Vasopressors (proportion)	18/22	16/20	$p > 0.05$
Thermal//Electrical burn (ratio)	20/2	16/4	$p > 0.05$
Biomarkers at ATBs TDM (CSLI)			
C- reactive protein (<0.5 mg/L)	208(115-282)	224(175-286)	$p > 0.05$
Leucocytes (4-11*1000 cells/mm3)	12.70(12.65-20.29)	11.23(10.97-18.53)	$p > 0.05$
Neutrophils (1.6-7*1000 cells/mm3)	9.42(6.66-13.45)	9.47(7.83-15.98)	$p > 0.05$
Creatinine clearance (90-120 mL/min)	121(87-159)	143(94-165)	$p > 0.05$

Abbreviations: ATBs: antibiotics; TDM: therapeutic drug monitoring; CSLI: Clinical Standards Laboratory Institute data base; IQ: quartiles (25-75); TBSA: total burn surface area; ICU: intensive care unit.

Statistics: Mann Whitney applied to unpaired population data, Fisher test, GraphPad Prism software, version 9.1.4.

Demographic, and clinical characteristics of patients at admission, and during drug serum monitoring in the Intensive Care Burn Unit (ICBU) are shown in Table 2; creatinine clearance was estimated by Schwartz's and compared to Cockcroft-Gault equation for G2-patients (14-18 yrs.).

Vancomycin Effectiveness Based on Serum Monitoring

Septic major pediatric burn patients 42 (31M/11F) with preserved or augmented renal function undergoing vancomycin therapy were investigated and distributed by age in two groups G1: 2-13 yrs. (n = 22) and G2:14-18 yrs. (n = 20). Vancomycin therapeutic target $AUC^{ss}_{0\text{-}24}/MIC > 400$ recommended at the last consensus, was considered in the study to investigate drug effectiveness by applying PK/PD target based on serum levels in pediatric major burn patients allocated by age (Rybak et al., 2020). Previous pharmacokinetic studies of dose adjustment done in pediatric or in adult septic burns, it is well known that its coverage is impacted by differences in pharmacokinetics of vancomycin, due a reduction on biological half-life due to increases on total body clearance age dependently, or even changes that occurs in drug elimination by vasopressors requirements were expected at the earlier stage of septic shock (Boeckh et al., 1988; Eiland et al., 2011; Elligsen et al., 2015; Frymoyer et al., 2009; Machado et al., 2007; Pires et al., 2020).

It was demonstrated in Set 1 that therapeutic target was attained up to MIC 1 mg/L against Gram-positive strains, only in 59% of patients, once coverage occurred in 13/22 G1- patients receiving the same empiric dose regimen (10-15 mg/kg q6h, 1 hr infusion) recommended for ICU pediatric patients. Finally, considering the initial dose regimen by 1 hr infusion (Set 1) recommended, the target was attained by vancomycin up to MIC 1 mg/L against Gram-positive strains only in 55% patients receiving dose regimen recommended 10-15 mg/kg q6h in G1-patients (13/22); 10-15 mg/kg q8h in G2-patients (6/20). While individualized therapy done in a real time, target attainment was achieved for all G1-patients' Gram-positive strains up to MIC 1 mg/L isolated from cultures; it is important to highlight that the coverage against MIC 2 mg/L was extended to 15/22 G1-patients and for all G2-patients. Therefore, individualization of drug therapy for vancomycin effectiveness, based on microbiology of isolates, was required up to MIC 1 mg/L for all patients, and it was extended to MIC 2 mg/L for 35/42 patients investigated. Consequently, based on this strategy, clinical and microbiological cure occurred of all patients in a short period, Table 3.

Table 3. Empirical and adjusted dose regimens. PK/PD approach for Vancomycin coverage clinical outcome against MIC 2 mg/L gram-positive strains isolates

Pediatric septic burn patients undergoing Vancomycin therapy, medians (quartiles, 25-75)			
Intermittent infusion 1 hr.	G1 n = 22(2-13 yrs)	G2 n = 20(14-18 yrs)	p
Empirical dose regimen	10-15 mg/kg q6h	10-15 mg/kg q8h	p < 0.05
Empirical daily dose (mg/kg)	40-60	30-45	p < 0.05
Adjusted dose regimen	20-39 mg/kg q6h	18-25 mg/kg q8h	p < 0.05
Adjusted daily dose (mg/kg)	80-156	54-75	p < 0.05
PK/PD approach	% Coverage	% Coverage	
Empirical dose regimen	Patients	Patients	p
Coverage MIC 0.5 mg/L	22/22 (100%)	0.5 mg/L (100%)	p > 0.05
Coverage MIC 1.0 mg/L	13/22 (45%)	1 mg/L (100%)	p < 0.05
Coverage MIC 2.0 mg/L	0/22 (0%)	2 mg/L (30%)	p < 0.05
Adjusted dose regimen	Patients	Patients	p
Coverage MIC 0.5 mg/L	22/22 (100%)	20/20 (100%)	p > 0.05
Coverage MIC 1.0 mg/L	22/22 (100%)	20/20 (100%)	p > 0.05
Coverage MIC 2.0 mg/L	15/22 (64%)	20/20 (100%)	p < 0.05
Clinical outcome	Period (days)	Period (days)	p
Negative Cultures	Up to 10	Up to 12	p > 0.05
ICU period (days)	23(19-31)	36(29-44)	p < 0.05
LOS in hospital(days)	28(29-34)	42(39-60)	p < 0.05
Survivals	Patients	Patients	p
Patients n = 42	22	20	p > 0.05

Abbreviations: LOS: length of stay; TDM: therapeutic drug monitoring; MIC: minimum inhibitory concentration.

Statistics: Mann Whitney applied to unpaired population data, Fisher test, GraphPad Prism software, v. 9.1.4.

Meropenem Effectiveness Based on Serum Monitoring

Pediatric major burns (42: 31M/11F) receiving meropenem and vasopressors at the earlier period of septic shock were investigated. All patients allocated by age were distributed in two groups. Dose regimen 40mg/kg q8h (120 mg/kg daily) done by 3 hrs.-infusion were considered for G1: 2-13 yrs. (n = 22) and dose regimen 20 mg/kg q8h (60 mg/kg/daily) eq. 1g q8h done by 3 hrs.-infusion was based on body weight prescribed to G2:14-18 yrs. (n = 20) to attain the target (100%$f\Delta T$ > MIC). Clinical and microbiological cure by meropenem occurred for all patients, once Gram-negative strains isolates up to MIC 2 mg/L were eradicated in a short period, considering MIC data of isolates from cultures in the Microbiology of central laboratory of hospital based on Clinical Standard Laboratory Institute database (CSLI, USA). Then, target was attained by meropenem for all patients up to MIC 4 mg/L against Gram-negative strains, Table 4.

Table 4. Empirical and adjusted dose regimen. PK/PD approach and outcome pediatric burn patients undergoing meropenem therapy of septic shock, medians (quartiles)

Pediatric patients	G1 n1 = 22 (2-13 yrs)	G2 n2 = 20 (14-18 yrs)	P
Extended 3 hrs- infusion	Meropenem	Meropenem	
Empirical dose regimen	40 mg/kg q8h	20 mg/kg q8h	$P < 0.05$
Daily dose (mg/kg)	120	60	$P < 0.05$
PK/PD approach	Patients	Patients	P
Coverage MIC 2 mg/L	22/22 (100%)	20/20 (100%)	$P > 0.05$
Coverage MIC 4 mg/L	22/22 (100%)	18/20 (90%)	$P > 0.05$
Coverage MIC 8 mg/L	12/22 (55%)	6/20 (30%)	$P < 0.05$
Clinical outcome	Period (days)	Period (days)	P
Negative Cultures	Up to 12	Up to 14	$P > 0.05$
ICU period (days)	22(18-28)	36(29-44)	$P < 0.05$
LOS in hospital(days)	25(23-32)	42(39-60)	$P < 0.05$
Survivals	Patients	Patients	P
Patients n = 42	22/22	20/20	$P > 0.05$

Abbreviation ICU: Intensive care unit; LOS: length of stay; MIC: minimum inhibitory concentration; quartiles: 25-75.

Statistics: Mann Whitney applied to unpaired population data, GraphPad Prism software, version 9.1.4.

Conclusion

Superiority of a new liquid chromatographic bioanalytical method developed and validated was shown based on larger linearity and lower limits of quantification and detection by 5 μL injected volume, with acceptable intra-day/inter day precision and systematic error. Good stability of standards on the bench (4 hrs.), both analytes in serum samples short stability on the bench (6 hrs.), post-process stability of serum extracts up to 24 hrs. in the rack in the autosampler, and serum samples long stability in ultralow freezer storage, as well as cycles freezing/de-freezing stability, and the robustness of chromatographic system.

In addition, vancomycin therapy was guided by cultures and PK/PD target attainment recommended AUC/MIC > 400-600 based on TDM of serum levels done in a real time. Vancomycin dose adjustment in septic pediatric burns receiving vasopressors was done if required against Gram-positive isolates (MIC 1-2 mg/L) for target attainment permitting an earlier clinical intervention to reach the desired the outcome in a short period. It was shown differences between groups related to vancomycin therapy due to age and vasopressor requirements. Then, dose adjustment done in a real time must be considered to guarantee drug effectiveness up to MIC 2 mg/L strains and cure of pediatric burn patient.

Concerning therapy of septic shock against Gram-negatives by meropenem 3 hrs.-infusion in pediatric septic burns, clinical and microbiological cure occurred for all patients from G1 (2-13yrs) receiving 40 mg/kg q8h and G2 (14-18 yrs.) at dose regimen 20 mg/kg q8h against strains up to MIC 4 mg/L. It is important to highlight that difference between groups was pointed out as a function of dose regimens, age, and vasopressors requirements.

Finally, PK/PD approach done in real-time based on drug serum monitoring by LC-UV is an important tool to assess soon drug effectiveness-safety in septic pediatric major burns. Consequently, combined antimicrobial therapy against Gram-positive and Gram-negative nosocomial pathogens for the combat of microbial resistance must be considered by applying bioanalytical methods based on liquid chromatography, that is quite selective for antimicrobial TDM serum levels done in a real time.

Disclaimer

None.

References

Abdul-Aziz M. H., Lipman, J., Mouton, J. W., Hope, W. W., Roberts, J. A. Applying pharmacokinetic/pharmacodynamic principles in critically ill patients: optimizing efficacy and reducing resistance development. *Semin Respir Crit Care Med.* 36(1):136–53,2015.

Abdul-Aziz M. H., Lipman J., Akova M., Bassetti M., De Waele J. J., Dimopoulos G., Dulhunty J., Kaukonen K. M., Koulenti D., Martin C., Montravers P., Rello J., Rhodes A., Starr T., Wallis S. C., Roberts J. A., DALI Study Group. Is prolonged infusion of piperacillin/tazobactam and meropenem in critically ill patients associated with improved pharmacokinetic/pharmacodynamic and patient outcomes? An observation from the Defining Antibiotic Levels in Intensive care unit patients. *DA. Journal of Antimicrobial Chemotherapy*, v.71, n.1, p.196–207,2016.

Abdulla A., Ewoldt T. M. J., Hunfeld N. G. M., Muller A. E., Rietdijk W. J. R., Polinder S., van Gelder T., Endeman H., Koch B. C. P. The effect of therapeutic drug monitoring of beta-lactam and fluoroquinolones on clinical outcome in critically ill patients: the DOLPHIN trial protocol of a multi-Centre randomized controlled trial. *BMC Infectious Diseases* (2020).

Barlam T. F., Cosgrove S. E., Abbo L. M., MacDougall C., Schuetz A. N., Septimus E. J., Srinivasan A., Dellit T. H., Falck-Ytter Y. T., Fishman N. O., Hamilton C. W., Jenkins T. C., Lipsett P. A., Malani P. N., May L. S., Moran G. J., Neuhauser M. M., Newland J. G., Ohl C. A., Samore M. H., Seo S. K., Trivedi K. K. Implementing an Antibiotic Stewardship Program: Guidelines by the Infectious Diseases Society of America and the Society for Healthcare Epidemiology of America. *Clinical Infectious Diseases*. v. 62. n. 10. p. e51–e77. 13 April 2016.

Boeckh M., Lode H., Borner K., Höffken G., Wagner J., Koeppe P. Pharmacokinetics and serum bactericidal activity of vancomycin alone and in combination with ceftazidime in healthy-volunteers. *Antimicrobial Agents and Chemotherapy*. 1988;32(1):92–95. predictability of vancomycin concentration using population pharmacokinetic parameters in pediatric patients. *Drug Metab Pharmacokinet*. 2005;20(6):415-422.

Cain S. E., Kohn J., Bookstaver P. B., Albrecht H., Al-Hasan M. N. Stratification of the Impact of Inappropriate Empirical Antimicrobial Therapy for Gram-Negative Bloodstream Infections by Predicted Prognosis. *Antimicrobial Agents and Chemotherapy*. 2015;59(1):245–250.

Cao Y., Yu J., Chen Y., Zhang J., Wu X., Zhang Y., Li G. Development and Validation of a New Ultra-Performance Liquid Chromatographic Method for Vancomycin Assay in Serum and Its Application to Therapeutic Drug Monitoring. *Ther. Drug Monit*. 2014, 36, 175–181.

Carlier M., Stove V., Wallis S. C., De Waele J. J., Verstraete A. G., Lipman J., Roberts J. A. Assays for Therapeutic Drug Monitoring of β-Lactam Antibiotics: A Structured Review. *International Journal of Antimicrobial Agents*. v. 46. n. 4. p. 367–375. October 2015.

Caro Y. S., Cámara M. S., De Zan M. M. A review of bioanalytical methods for therapeutic drug monitoring of beta-lactam antibiotics in critically ill patients: Evaluation of the approaches used to develop and validate quality attributes. *Talanta* 210(2020) 120619.

Casals G., Hernández C., Hidalgo S., Morales B., López-Púa Y., Castro P., Fortuna V., Martínez J. A., Brunet M. Development and validation of a UHPLC diode array detector method for meropenem quantification in human plasma. *Clin. Biochem.* 47 (2014) 223–227.

Crossley K. B., Rotschafer J. C., Chern M. M., Mead K. E., Zaske D. E. Comparison of a Radioimmunoassay, and a Microbiological Assay for Measurement of Serum Vancomycin Concentrations. *AAC* Apr. 1980, p. 654-657, v. 17, No.4.

Demotes-Mainard F., Labat L., Vinçon G., Bannwarth B. Column-Switching High-Performance liquid Chromatographic Determination of Vancomycin in Serum. *Ther. Drug Monit.* 1994, 16, 293–297.

Eiland L., English T. and Eiland E. Assessment of Vancomycin Dosing and Subsequent Serum Concentrations in Pediatric Patients. *Annals of Pharmacotherapy*. 2011;45(5):582–589.

Elligsen M., Walker S. A. N., Simor, A. Optimizing Initial Vancomycin Dosing in Burn Patients. *Burns.* v. 37. n. 3. p. 406–414. May 2011.

Evans L., Rhodes A., Alhazzani W., Antonelli M., Coopersmith C. M., French C., Machado F. R., Mcintyre L., Ostermann M., Prescott H. C., Schorr C., Simpson S., Wiersinga W. J., Alshamsi F., Angus D. C., Arabi Y., Azevedo L., Beale R., Beilman G., Belley-Cote E., Burry L., Cecconi M., Centofanti J., Coz Yataco A., De Waele J., Dellinger R. P., Doi K., Du B., Estenssoro E., Ferrer R., Gomersall C., Hodgson C., Møller M. H., Iwashyna T., Jacob S., Kleinpell R., Klompas M., Koh Y., Kumar A., Kwizera A., Lobo S., Masur H., McGloughlin S., Mehta S., Mehta Y., Mer M., Nunnally M., Oczkowski S., Osborn T., Papathanassoglou E., Perner A., Puskarich M., Roberts J., Schweickert W., Seckel M., Sevransky J., Sprung C. L., Welte T., Zimmerman J., Levy M. Surviving Sepsis Campaign: International Guidelines for Management of Sepsis and Septic Shock 2021, *Critical Care Medicine*: November 2021 - Volume 49 – Issue 11 - p 1063-1143.

Ferrone V., Cotellese R., Di Marco L., Bacchi S., Carlucci M., Cichella A., Raimondi P., Carlucci G. Meropenem, levofloxacin and linezolid in human plasma of critical care patients: a fast semi-automated micro-extraction by packed sorbent UHPLC-PDA method for their simultaneous determination, *J. Pharm. Biomed. Anal.* 140 (2017) 266–273.

Fournier A., Eggimann P., Pagani J. L., Revelly J. P., Decosterd L. A., Marchetti O., Pannatier A., Voirol P., Que Y. A. Impact of the introduction of real-time therapeutic drug monitoring on empirical doses of carbapenems in critically ill burn patient. *Burns* 41 (2015).

Fridlund J., Woksepp H., Schön T. A microbiological method for determining serum levels of broad spectrum β-lactam antibiotics in critically ill patients. *Journal of Microbiological Methods* 129 (2016) 23-27.

Frymoyer A., Hersh A. L., Benet L. Z., Guglielmo B. J. Current Recommended Dosing of Vancomycin for Children with Invasive Methicillin-Resistant Staphylococcus aureus Infections Is Inadequate. *Pediatric Infectious Disease Journal.* 2009;28(5):398–402.

Furuta I., Kitahashi T., Kuroda T., Nishio H., Oka C., Morishima Y. Rapid serum vancomycin assay by high-performance liquid chromatography using a semipermeable surface packing material column. *Clin. Chim. Acta* 2000, 301, 31–39.

Global Sepsis Alliance. 4th World Sepsis Congress on April 25 and 26, 2023. One Global Health Treat: Sepsis, Pandemics, and Antimicrobial Resistance. Available in: https://www.worldsepsiscongress.org.

Global Sepsis Alliance. Sepsis and COVID-19/Coronavirus/SARS-COV-2, 2020. Available in: https://www.global-sepsis-alliance.org/covid19.

Greenhalgh D. G., Saffle J. R., Holmes J. H. 4th, Gamelli R. L., Palmieri T. L., Horton J. W., Tompkins R. G., Traber D. L., Mozingo D. W., Deitch E. A., Goodwin C. W., Herndon D. N., Gallagher J. J., Sanford A. P., Jeng J. C., Ahrenholz D. H., Neely A. N., O'Mara M. S., Wolf S. E., Purdue G. F., Garner W. L., Yowler C. J., Latenser B. A. American burn association consensus conference to define sepsis and infection in burns. *Journal of Burn Care & Research.* 2007;28(6):776–790.

Hagihara M., Sutherland C., Nicolau D. P. Development of HPLC Methods for the Determination of Vancomycin in Human Plasma, Mouse Serum, and Bronchoalveolar Lavage Fluid. *Chromatogr. Sci*. 2013,51:201–207.

Javorska L., Krcmova L. K., Solichova D., Solich P., Kaska M. Modern methods for vancomycin determination in biological fluids by methods based on high performance liquid chromatography – *A review. Sep. Sci*. 2016, 39, 6-20.

López K. J., Bertoluci D. F., Vicente K. M., Dell'Aquilla A. M., Santos S. R. Simultaneous determination of cefepime, vancomycin and imipenem in human plasma of burn patients by high-performance liquid chromatography. *Journal of Chromatography B-Analytical Technologies in the Biomedical and Life Sciences*. 2007;860(2):241–245.

Machado J. K., Feferbaum R., Kobayashi C. E., Sanches C., Santos S. R. Vancomycin pharmacokinetics in preterm infants. *Clinics*. 2007;62(4):405–410.

McWhinney B. C., Wallis S. C., Hillister T., Roberts J. A., Lipman J., Ungerer J. P. Analysis of 12 beta-lactam antibiotics in human plasma by HPLC with ultraviolet detection. *J. Chromatogr*. B 878 (2010) 2039–2043.

Pires F. R., Paula S. I., Delgado A. F., Carvalho W. B., Duarte N. J. C., Morales Júnior R., Santos S. R. C. J. Does vancomycin administered at an empirical dose ensure coverage of pediatric patients against gram positive pathogens? *Brazilian Journal of Intensive Therapy*. v.32(3),391-397. ISSN 0103-507x.

Roberts J. A., Hope W. W., Lipman J. Therapeutic Drug Monitoring of β-Lactams for Critically Ill Patients: Unwarranted or Essential? *International Journal of Antimicrobial Agents*. v. 35. n. 5. p. 419–420. May 2010.

Roth T., Fiedler S., Mihai S., Parsch H. Determination of meropenem levels in human serum by high-performance liquid chromatography with ultraviolet detection. *Biomed. Chromatogr*. (2017) 31.

Rudd K. E., Johnson S. C., Agesa K. M., Shackelford K. A., Tsoi D., Kievlan D. R., Colombara D. V., Ikuta K. S., Kissoon N., Finfer S., Fleischmann-Struzek C., Machado F. R., Reinhart K. K., Rowan K., Seymour C. W., Watson R. S., West T. E., Marinho F., Hay S. I., Lozano R., Lopez A. D., Angus D. C., Murray C. J. L., Naghavi M. Global. Regional and National Sepsis Incidence and Mortality. 1990–2017: Analysis for the Global Burden of Disease Study. *The Lancet*. v. 395. n. 10219. p. 200–211. jan. 2020.

Rybak M. J., Le J.., Lodise T. P., Levine D. P., Bradley J. S., Liu C., Mueller B. A., Pai M. P., Wong-Beringer A., Rotschafer J. C., Rodvold K. A., Maples H. D., Lomaestro B. M. Therapeutic monitoring of vancomycin for serious methicillin-resistant Staphylococcus aureus infections: A revised consensus guideline and review by the American Society of Health-System Pharmacists, the Infectious Diseases Society of America, the Pediatric Infectious Diseases Society, and the Society of Infectious Diseases Pharmacists. *Am J Health Syst Pharm*. 2020;77(11):835–864.

Santos SRCJ, Sanches-Giraud C, Silva Júnior CV, Gomez DS. Pharmacokinetic-pharmacodynamic correlation for meropenem applied to a burn child using a bioanalytical liquid chromatographic method. *Revista Portuguesa de Farmacoterapia*. 2011;3(4):224–232.

Schwartz G. J., Haycock G. B., Edelmann C. M. Jr, Spitzer A. A simple estimate of glomerular filtration rate in children derived from body length and plasma creatinine. *Pediatrics.* 1976;58(2):259–263.

World Health Organization. From emergency response to long-term covid-19 disease management: Sustaining gains made during the COVID-19 pandemic. May 4th, 2023. Available in: https://www.who.int/publications/i/item/WHO-WHE-SPP-2023.1.

Zou L., Meng F., Hu L., Huang Q., Liu M., Yin T. A novel reversed-phase high-performance liquid chromatographic assay for the simultaneous determination of imipenem and meropenem in human plasma and its application in TDM. *J. Pharm. Biomed. Anal.* 169 (2019) 142–150.

Biographical Sketch

Frederico Ribeiro Pires

Affiliation:

University Federal of Brasilia – Medical School (2010), Intensive Care Pediatric Unit Residence (2011-2014), Expertise in Intensive Care in Pediatric Intensive Care Unit.

Education:

University of São Paulo Medical Doctor_Expertise in Intensive Care in Pediatric Intensive Care Unit (2023).

Research Experience:

Patients in hospital ICU of Burn and Pediatric Department of Pediatrics – Medical School of University of São Paulo (2016-23).

Professional Experience:

Antimicrobial effectiveness and safety by individualization of therapy in ICU critically ill pediatric patients by therapeutic drug monitoring and PK/PD approach based on serum levels at the Clinical Pharmacokinetics Center – University of Sao Paulo.

Professional Appointments:

Research fellow of team on Clinical Pharmacokinetics and Medical doctor -Pediatrics Intensive Care Units Federal hospital of Brasilia, DF (2023).

Honors:

III International Meeting Sobratafe on Advanced Wound Care, Sobratfe 2019, Maksoud Plaza Hotel São Paulo, SP. Awards At The Event Promoter: Brazilian Society for Advanced Treatment of Complex Wounds (Sobratafe)

"Clinical Protocol" Award 2: Vancomycin dose adjustment done in areal time for target attainment in critically ill septic pediatric burns versus non-burns against Staphylococcus spp MIC 1 mg/L. Camila Salinas Guillaux, Ronaldo Morales Júnior, Vedilaine Aparecida Bueno da Silva Macedo, Frederico Ribeiro Pires, Elson Mendes da Silva Júnior, João Manoel da Silva Júnior, Edvaldo Vieira de Campos, Amanda Maria Ribas Rosa de Oliveira, Aline Sandré Gomides, Gabriela Aparecida Ferreira, Thiago C. Oliveira, Silvia Regina Cavani Jorge Santos, David de Souza Gomez.

Research team member since 2016

Publications from the Last 3 Years:

1) Morales, Jr, R., Macedo, V. A. S., Pires, F. R., Campos, E. V., Ebner, P. A. R., Romano, P., Maria Severina dos Santos, Duarte, N. J., Gomez, D. S. & Santos, S. R. C. J. Vancomycin target attainment in critically ill septic pediatric burn versus non-burn patients against Staphylococcus spp. *International Journal of Antimicrobial Agents*, 58, 2100282. https://doi.org/10.1016/j.ijantimicag.2021.106421.11.
2) de Camargo, T. V., Junior, E. M. S., Silva Jr, J. M., Gomez, D. S. & Santos, S. R. C. J. (2022). PK/PD approach to evaluate Meropenem effectiveness in critically ill burn adolescents versus young adults undergoing therapy of septic shock. *Pharm Pharmacol Int J*, 10(3), 79-85 (2022). https://doi.org/10.15406/ppij.2022.10.00368.
3) Santos, S. R. C. J., de Camargo T. V., Macedo V. A., Souza, V. K. & Gomez, D. S. Vancomycin and meropenem serum monitoring for target attainment by PK/PD approach as an effective tool in the battle against nosocomial pathogens in septic pediatric burn patients. *Pharm Pharmacol Int J.* 2022;10(3):87–95 (2022). https://doi.org/10.15406/ppij.2022.10.00369.
4) Santos, S. R. C. J., de Camargo T. V., Messiano C. G., Kupa L. V. K., Souza V. K., Ronaldo Morales Jr. R., Sanches Pinto D. C. S., Silva Junior E. M., Silva Junior J. M., Gomez D. S. Combating bacterial resistance to antimicrobials in severe septic ICU patients: importance of

meropenem, piperacillin serum monitoring as a dose adjustment and duration of infusion strategies. *Pharm Pharmacol Int J.* 2023;11(2):52–60 (2023). https://doi.org/10.15406/ppij.2023.11.00403.

5) Santos, S. R. C. J., de Camargo T. V., Messiano C. G., Souza V. K. Frederico Ribeiro Pires, Débora Cristina Sanches Pinto, Edvaldo Vieira de Campos, Elson Mendes da Silva Junior, João Manoel da Silva Junior, David de Souza Gomez. Combined vancomycin-meropenem therapy in pediatric major burns undergoing therapy of septic shock guided by cultures and pharmacokinetic-pharmacodynamics approach based on serum levels to combat bacterial resistance. *Pharm Pharmacol Int J.* 2023;11(2):62–68 (2023). https://doi.org/10.15406/ppij.2023.11.00402.

Chapter 4

Liquid Chromatography-Tandem Mass Spectrometry for Simultaneous Piperacillin-Meropenem Serum Measurements in Septic Burns for Effectiveness

Paschoalina Romano[1], PhD
Pérsio de Almeida Rezende Ebner[1], BS
Marc Yves Chalom[1], BS
Maria Severina dos Santos[1], BS
Marcio Santos Garcia[1], BS
Kupa Leonard de Vinci Kuppa Kanda[2], PhD
Nairo Massakazu Sumita[1], PhD
Nilo José Coêlho Duarte[1], MD
Aline Sandre[3], BS
Gabriela Aparecida Ferreira[3], BS
Joao Manoel da Silva Junior[3], MD
Thais Vieira de Camargo[3], PhMS
Ronaldo Morales Junior[3]
David de Souza Gomez[3], MD
and Silvia Regina Cavani Jorge Santos[3,*], PhD

[*] Corresponding Author's Email: pharther@usp.br.

In: Chromatographic Methods and Research
Editor: Raven W. Obrien
ISBN: 979-8-89113-193-4

[1]Central Laboratory Division and Medical Investigation Laboratory (LIM 03) – Medical School, University of Sao Paulo (HCFMUSP).
[2]Medical investigation Laboratory (LIM 17), Medical School, University of Sao Paulo, Sao Paulo, Brazil
[3]Division of Plastic Surgery and Burns, Clinics Hospital Department of Surgery, Medical School, University of Sao Paulo, Sao Paulo/SP, Brazil

Abstract

Meropenem and Piperacillin-Tazobactam are routinely prescribed to critically ill septic patients against infections caused by Gram-negative nosocomial pathogens. Then, therapeutic drug monitoring (TDM) based on serum levels of these antibiotics, pharmacokinetics and pharmacokinetic-pharmacodynamics approach are required to reduce mortality of those patients undergoing intensive care. Consequently, liquid chromatography-tandem mass spectrometry (UHLCMS/MS) was chosen for simultaneous drug serum measurements in a public tertiary hospital, due to its high specificity and selectivity, once the technique is considering the gold standard for antibiotics TDM. It was developed and validated a new bioanalytical method by applying UHLCMS/MS technique to quantify meropenem and piperacillin in human serum for comparison purposes with a conventional LC-UV technique routinely applied previously for these antibiotics serum measurements in septic burn patients from the Intensive Care Unit (ICU) of a tertiary public hospital. The SRM quantitation transitions selected for positive electrospray ionization was used in a triple quadrupole LC-MS/MS were m/z 384.1 > 141.1 for meropenem (MEP) and 390.1 > 146.1 for MEP-d6 (IS) and 535.2 > 143.1 for piperacillin (PIP) and 540.3 > 148.1 for PIP-d5 (IS). To each 100 µL of serum sample, 10 µL (20 µg/mL) of internal standard mixture was added, followed by 500 µL acetonitrile. Samples were centrifuged at 8,000 rpm for 10 min. Supernatant was diluted with deionized water (1:1, v/v) and 10 µL were injected to a Hypersil Accucore C18 column (100 mm x 2.1 mm id, 2.6 microns). Elution with a linear gradient was done at a flow rate of 0.4 mL/min with acetonitrile containing 0.1% formic acid, and 2 mmol /L ammonium acetate with 0,1% formic acid. Validation of the method was based on the international standards. It was shown linearity from 1.0 up to 250 µg/mL ($r^2 > 0.995$). A lower limit of quantification (LLOQ) of 1.0 µg/mL was established for both analytes; detection limits of 0.27 µg/mL for meropenem, and 0.24 µg/mL for piperacillin were found. Precision of 7.5-8.6% for meropenem and 5.5-7.4% for piperacillin at low, medium, and high serum levels were considered. Relative recovery ranged 87.8 – 114.2% for meropenem, and 93.7-118.2% for piperacillin. Clinical study was designed with twenty-eight adult ICU major burn patients undergoing therapy of septic shock; patients included were distributed in two groups based on cultures and antibiotic prescription chosen. Routinely, only in this ICU of hospital, an extended 3-hrs.- infusion was chosen for these major burn septic patients at dose regimens recommended to meropenem, 1g q8h (G1: n = 12 patients), or piperacillin-tazobactam of 4.5g q6h (G2: n = 16 patients). Blood was sampling after 48 hours of therapy started, mainly at the early phase of septic shock, and only two blood samples were collected (4 mL/each) at the 3rd of the beginning of infusion, and one hour before the next infusion. Serum samples from patients, who received meropenem or piperacillin-tazobactam were analyzed by both methods based on liquid chromatography procedures (LC-UV *versus* UPLCMS/MS). It was shown a very good correlation between chromatographic techniques ($r^2 = 0.9988$ $p = 0.766$, $p > 0.05$) for meropenem analysis. Also, a very good correlation between chromatographic techniques ($r^2 = 0.9620$ $p = 0.457$, $p>0.05$) was shown for piperacillin serum samples. Clinical cure occurred for all meropenem G1-12 patients by eradicating eight isolate pathogens susceptible up to MIC 2 mg/L strains, as well as two isolates MIC 4 mg/L, intermediate susceptibility of *P. aeruginosa* and *K. pneumoniae.* Clinical cure occurred also for all G2-16 patients by piperacillin-tazobactam against 19 Gram-negative susceptible strains isolated from cultures up to MIC 16 mg/L. UPLCMS/MS method was successfully developed and validated, with run time of lower than five minutes per sample analyzed overnight in the autosampler. Advantage of UPLCMS/MS is an opportunity for the simultaneous serum measurements of multiple β-lactam antibiotics with high selectivity, specificity, simple extraction procedure, short run time, reduced serum volume requirements.

Keywords: UHLCMS/MS serum levels, meropenem-piperacillin therapeutic serum monitoring, PK/PD approach in septic burns, drug effectiveness by 3hrs.-infusion against Gram-negative strains

Abbreviations

FDA	Food Drug Administration
GSA	Global Sepsis Alliance
ICU	Intensive Care Unit
MDR	Multidrug Resistance
MIC	Minimum Inhibitory Concentration
MV`	Mechanical Ventilation

PK/PD	Pharmacokinetics- Pharmacodynamics
PTA	Probability of Target Attainment
SAPS3	Simplified Acute Physiology Score 3
SARS-	Severe Acute Respiratory Syndrome Coronavirus 2 CoV-2
SIRS	Systemic Inflammatory Response Syndrome
TBSA	Total Burn Surface Area
TDM	Therapeutic Drug Monitoring
WHO	World Health Organization

Introduction

Meropenem and Piperacillin/tazobactam is prescribed to critically ill septic patients with nosocomial infections caused by Gram-negative strains. Meropenem and Piperacillin/tazobactam is prescribed to critically ill septic patients with nosocomial infections caused by Gram-negative strains. Infection is a serious concern among hospitalized patients and require innovative pharmacokinetic dosing strategies to achieve clinical success, especially as the emergence of resistant Gram-negative pathogens has outpaced the development of new antibiotics (World Health Organization. 2023; Global Sepsis Alliance, 2023). Beta-lactam antibiotics exhibit time-dependent activity, meaning that optimal efficacy is achieved when free drug concentrations remain above the minimum inhibitory concentration for an extended period of the recommended dose regimens and a predicted prognosis related to the impact of inappropriate empirical antimicrobial therapy for Gram-negative bloodstream infections can be done (Abdul-Aziz et al., 2015; Cain et al., 2014). On the other hand, the success on therapeutic rate with first, second, third and fourth cephalosporin generations was demonstrated in patients with febrile neutropenia by reducing the time on mechanical ventilation, since in general ventilator-associated pneumonia as caused by multidrug-resistant pathogens (Karim et al., 2022; Sime et al., 2015).

Meropenem and piperacillin-tazobactam therapies by extended infusion has been demonstrated by improvement of clinical outcome by decreasing mortality in critically ill patients with Gram-negative infections caused specially by Enterobacteriaceae and *Pseudomonas aeruginosa* MIC 16 mg/L, a non-Enterobacteriaceae of high incidence and prevalence in the most ICUs (Barlam et al., 2016; Global Sepsis Alliance, 2020; Rhodes et al., 2017; Rudd et al., 2020). Additionally, the extended infusion strategy of 3-4 hours for meropenem or even for piperacillin brings additional benefits related to increased antimicrobial effectiveness, reducing the length of stay in the ICU, and consequently also including a significant reduction of hospital costs related to prolonged hospitalization (Abdul-Aziz et al. 2016; Abdulla et al., 2020; Chung et al., 2015; Evans et al., 2021, Fournier et al., 2015). In contrast, it is important to highlight that in a recently published study, continuous infusion of meropenem was compared with intermittent infusion. Thus, 607 included patients were investigated, 61% with septic shock for the measurement of the 28-day primary outcome and completed the 90-day mortality follow-up.

In conclusion, critically ill patients with sepsis, intermittent infusion compared to continuous infusion of meropenem did not improve the outcome of mortality and emergence of pan-resistant or extensively drug-resistant bacteria on day 28 (Monti et al., 2023). Pathophysiologic changes, such as pharmacokinetics alterations in drug clearance and protein binding, volume of distribution and acid-base balance, which alter drug distribution and elimination in these ICU patients during the septic shock, can significantly affect the pharmacokinetics (PK) profile of prescribed antibiotic, impacting the coverage (Gonçalves-Pereira et al., 2014). On the other hand, patients undergoing therapy of septic shock with hydrophilic beta lactam agents, as meropenem or piperacillin-tazobactam at recommended dose regimens done by intermittent or even by extended infusions may have augmented total body clearance by vasopressors with a consequent reduction on serum levels negatively impacting the desired outcome (Udy et al., 2015). Recently, it was demonstrated in a review study that the pharmacokinetics of two beta-lactam agents widely prescribed for critically ill ICU patients will depend on the duration of the infusion for meropenem (1g every 8h) or piperacillin-tazobactam (4.5g every 6h) performed by the intermittent infusion of 0.5 hr. or by the prolonged infusion of 3-4 hrs.

In that study, it is pointed out that the superiority of extended infusion over the intermittent infusion occurred for both antimicrobial agents in a different manner with an important positive impact on coverage investigated by PK/PD approach based on meropenem or piperacillin serum levels (Santos et al., 2023a). In addition, data reported recently by many authors indicate that changes on pharmacokinetics of both drugs occurs by cytokines storm because of the systemic inflammatory response syndrome (SIRS), that could result in serum levels lower than those required to achieve the therapeutic target for pathogen eradication, and the expected clinical outcome. Therefore, when critically ill patients, with preserved, or augmented renal clearance by vasopressor requirements, undergoing antimicrobial therapy by the extended

3 hrs.-infusion defined against Gram negative susceptible strains, the PK/PD target $100f\Delta T$ > MIC was attained for adult or pediatric septic patients up to MIC 2 mg/L by the meropenem, and extended up to MIC 4 - 8 mg/L strains of intermediate susceptibility. Considering piperacillin-tazobactam, coverage occurred against susceptible strains up to MIC 16 mg/L by piperacillin (CLSI database, USA). The PK/PD approach based on serum levels after blood sampling that was done for both antimicrobial agents, in early-stage septic shock, for the therapeutic drug monitoring purposes, permitted the assessment for antimicrobials coverage against Gram-negative strains isolated from cultures. (Camargo et al., 2022; Gonçalves-Pereira et al., 2014; Messiano et al., 2022; Santos et al., 2022; Taccone et al., 2010, Udy et al., 2015). On the other hand, ICU patients who have multiple failure of vital organs, such as renal and hepatic failure, which represent the main routes of elimination of antimicrobials prescribed in ICU septic patients, are at greater risk of toxicity due to increased serum levels. So, antimicrobial serum levels followed by the PK/PD approach are important tools and a good laboratory strategy to assess risk-benefit in these patients. It was evidenced that different antibiotics are affected in a different manner by the pathophysiological changes found in ICU septic patients. Then, the serum levels of them are routinely recommended in ICU patients to evaluate if the PK/PD target was attained to reach the clinical cure and desired clinical outcome. (Abdul-Aziz et al., 2016; Carlier et al., 2015; Fournier et al., 2015, Gonçalves-Pereira et al., 2014; Santos et al., 2023b; Udy et al., 2015).

Rational of Study

Subject of study was to develop and to validate a bioanalytical method for simultaneous meropenem and piperacillin serum measurements by UPLC-MS/MS technique to investigate meropenem and piperacillin-tazobactam effectiveness in recommended dose regimens using the strategy of an extended 3 hrs.-infusion by PK/PD approach.

Bioanalytical Method

Clinical study was designed to investigate the application of ultra-high performance liquid chromatography (UPLC) equipped with mass spectrometry detection (MS/MS) for meropenem and piperacillin serum measurements to investigate antimicrobial target attainment by PK/PD approach. Furthermore, a very good review with comparison of data related to chromatographic techniques has been described recently (Caro et al., 2020). So, the superiority of UHPLC-tandem mass spectrometry over the conventional HPLC technique by using various detectors such as ultraviolet, fluorescence and diode array was highlighted by these authors. Twenty-eight septic major burn patients, adults of both genders, with renal function preserved undergoing the recommended therapy with meropenem 1g q8h, or piperacillin-tazobactam 4.5g q6h by extended 3 hrs. infusion were investigated. Monitoring of cultures were done by blood collection, fluids and secretions prior to the antimicrobial therapy starts. Susceptibility testing for Gram-positive and Gram-negative were based on CSLI database in the Microbiology Laboratory of hospital. After 48 hrs. of antimicrobial therapy started, blood was sampling, 4mL/each, for antimicrobial serum measurements at the 3rd hr. of infusion started, and one hour before the next infusion. Serum samples (100μL) were analyzed LC-MS/MS by adding internal standard 10 μL (mix 20μg/mL: meropenem D6 and piperacillin D5) followed by deproteinization with acetonitrile (mass spectrometer grade) 500 μL, vortexed and centrifuged at 13,500 rpm for 8 minutes. Supernatant was diluted with deionized water (1:1, v/v) and 10 μL were injected into LC-MS/MS TSQ Acquity, Waters. Co. Separation was done in a column Hypersil Accucore C18 (100 x 2.1mm, id 2.6μm) in a gradient elution with mobile phase A: ammonium acetate 2mM in water and phase B: acetonitrile grade LCMS, both with 0.1% of formic acid added. Peaks were monitored at 0.4 mL/min of flow rate. Mass Spectrometer Settings: electrospray positive ionization; MRM positive mode; ESI voltage: 2.95kV; cone voltage: 36V; dessolvation temperature: 200°C. MRM Transition settings: Piperacillin: 535.2 > 143.1 (quantification); 535.2 > 359.1 (confirmation) Piperacillin D5: 540.3 > 148.1 (quantification); 540.3 > 364.1 (confirmation). Instrument was calibrated separately by using calibrators and samples with internal standards (IS) of two different matrices of the study (solvent and serum). Six different concentrations were used to obtain calibration curves (0-250 μg/mL) by linear regression of the peak area ratios of meropenem and piperacillin adjusted to the internal standard versus concentration. Calibration curve was obtained with the same matrix for analyses of patients' serum samples. Calibration curves were accepted based on systematic error lower than 15%, also for the internal controls (high, medium, and low serum concentrations); then, drug serum levels in patients' samples were determined based on the accepted calibration curve with the internal standard, ion ratio and sign/noise is calculated for each sample. The results are within the total acceptable error adopted for new drugs in our laboratory (25%) within the limits of acceptability established in the literature for precision, accuracy, lower limit of quantification and higher limit of quantification described previously (Lunn 2005, 234; Oliveira and Mendes, 2011). Precision analysis included replicates (n = 20) of two serum levels. Intra-assay precision was determined by repeated analysis performed two times in the same day, same system technician. Inter-assay precision was determined by repeated analysis performed on five consecutive days and for quantifications in a day at each concentration. Precision was calculated and expressed as the coefficient of variation among quantification in the same day (intra–assay) and between different days (inter-assay). A limit coefficient of variation lower than 15% was considered appropriate. Accuracy was estimated as the percentage of deviation between nominal and measured concentrations. It was considered adequate if around 15%. Linearity was assessed using five different calibrators in two replicates for the three different matrices studied. Quantification was considered linear if $r2 \geq 0.98$, and the systematic error $\leq 15\%$. Additionally, the calibrator of 250 µg/mL were analyzed in 20 replicates to verify the highest limit of quantification. A coefficient of variation limit lower than 15% was considered acceptable, except at the lower limit of quantification (LLOQ), that it should be up to 20%. Recovery with addition of standard samples was calculated as the percentage difference between meropenem or piperacillin quantification and their nominal amount. Matrix effect, carryover was also tested, and any interference of endogenous, exogenous compounds were registered (D'Cunha 2018, 15). Bioanalytical method was validated according to the Clinical Laboratorial Standards Institute (CLSI-C50A 2007; CLSI-EP6 2009; CLSI-EP 15A3 2014; CLSI-EP 62A) and FDA Guidelines (FDA 2020).

Clinical Trial Study: Study Design, Patient Eligibility, Meropenem and Piperacillin Therapy

The clinical protocol was a prospective, open-label study. Ethical register CAAE 07525118.3.0000.0068, Brazilian Platform was obtained by approval of the Ethical Committee of Hospital of Clinics, Medical School of University of Sao Paulo; no conflicts of interest to declare were obtained from authors. Study was conducted from January 2018 to Jully 2019; the informed written consent was obtained from all representative person related to each patient. The chosen therapy with meropenem or piperacillin were prescribed at dose regimens recommended for adult patients with renal function preserved or augmented by vasopressors requirements, 48 hours of therapy started. Septic major burn adult patients, at the earlier stage of septic shock, were included to investigate antimicrobial effectiveness of the therapy guided by PK/PD tools and cultures. Dose regimen of 1g q8h to G1-patients was related to meropenem therapy; while piperacillin-tazobactam was given to G2-patients, 4.5 g q6h, both by pump extended 3 hrs.-infusion at the dose regimen recommended, body weight normalized, according to the institutional protocol. Complete medical histories, physical examinations were obtained for each enrolled patient; laboratory data included documented strains in blood, bone, pulmonary lavage, and urine cultures. Susceptibility testing done in the Microbiology of central laboratory of hospital to obtain the minimum inhibitory concentration for each antimicrobial agent against each pathogen isolated. Blood was sampling at the 3rd hr. of the infusion started and one hour before the next infusion for antimicrobial serum levels. One compartment open model was chosen to investigate pharmacokinetics, and noncompartmental data analysis was applied. Drug effectiveness was evaluated by PK/PD approach based on target 100%$f\Delta T$ > MIC and serum levels.

Analytical Procedures

A good performance was obtained by development and validation of the bioanalytical method by applying UPLCMS/MS and results are within the total acceptable error adopted, within the limits established for precision, accuracy. Lower and upper limits of quantification were considered acceptable. It was detailed in Table 1 data from the validation study.

New bioanalytical method developed and validated by the UPLCMS/MS technique described was compared with the conventional technique HPLC-UV detection. It was shown a good correlation between chromatographic techniques ($r2 = 0.9988$ $p = 0.766$, $p > 0.05$) for meropenem analyzed in patients' serum samples, and for piperacillin serum samples ($r2 = 0.9620$ $p = 0.457$, $p > 0.05$), with correlation between both techniques. Consequently, a very good correlation was obtained based on data of all serum samples analyzed by HPLC-UV and UPLCMS/MS techniques, once any statistically significant differences between chromatographic techniques was shown.

Table 1. Bioanalytical method by liquid chromatography-tandem mass spectrometry for simultaneous meropenem-piperacillin serum monitoring

Parameters	Meropenem	Piperacillin
Linearity (lower/upper limits) µg/mL	1.0-246.1	1.0-237.6
Linear regression coefficient	0.9988	0.9970
LOOQ µg/mL	1.0	1.0
LOD µg/mL	0.27	0.24
Relative recovery (mean, CI95%)	99.1(87.8-114.2)	108.1(93.1-118.7)
Intraday precision (CV%) n = 20	7.2(5.2-7.3)	7.3(7.0-7.5)
Interday precision (CV%) n = 20	8.7(8.0- 9.4)	9.1(8.9-9.5)
Intra-day (accuracy) n = 20 (%)	107.6(97.0-115.3)	104.0(98.1-110.8)
Interday (accuracy) n = 20 (%)	105.5(95.7-119.0)	92.9(82.6-110.3)
Standard stability (8 h./30°C) n = 10 (%)	7.3(6.2-10.1)	6.7(5.6-13.2)
IS stability (8 h./30 °C) n = 10 (%)	4.5(3.0-4.8)	4.0(3.2-4.2)
Short stability (1day; 3cycles) n = 10 (%)	5.7(2.8-6.9)	8.0(3.0-8.0)
Post-P stability (7 days/30 °C) n = 10 (%)	7.0(4.7- 11.2)	9.9(8.7-10.9)
Long stability (30 days; -80°C) n = 20 (%)	8.0(7.5-9.9)	7.5(6.5-9.9)
Cycles de-freezing stability n = 10 (%)	9.0(5.5-9.9)	8.9(5.7-10.0)
Robustness (LC system) n = 20 (%)	6.3(8.2-9.2)	8.8(7.2-10.0)
Matrix effect (solvent/serum) n = 20 (%)	8.5(1.2-9.0)	5.1(1.8-8.9)
Carryover n = 21	- 0.32 (limit error = 0.37)	0.08 (limit error = 0.13)
Methods comparison (LC-UV *versus* UPLCMS/MS)	$r^2 = 0.999$ ($p = 0.766$)	$r^2 = 0.962$ ($p = 0.457$)

Abbreviations: CV: coefficient of variation; LOOQ: lower limit of quantification; LOD: limit of detection; CI: confidence interval; IS: internal standard; Post-P: post-process.

Statistics: data expressed by means (CI95%), GraphPad Prism v.9.4.1 software. significance: $p < 0.05$; $p > 0.05$ (not significant difference). GraphPad Prism v.9.4.1 software.

Clinical Study Meropenem and Piperacillin

Septic burn patients undergoing piperacillin-tazobactam therapy with a dose regimen of 4.5g q6h (n = 16) or receiving meropenem 1g q8h (n = 12) were done both by extended 3 hrs.- infusion. PK/PD target recommended of 100% $f\Delta T$ > CIM was considered. No statistical differences between groups were found related to the demographic data as gender, age, body weight and height, once a population of young adults with preserved or augmented renal clearance by vasopressor requirements was considered. On the other hand, it was found significant differences between groups at admission data related to total body surface area, SAPS3 (simplified acute physiologic score) and death risk score associated. In addition, no significant differences between groups (G1/G2) were found related to inhalation injury, thermal burn, mechanical ventilation, and vasopressors requirements. Then, meropenem was prescribed to G1-patients as a function of admission data related to TBSA > 40%, SAPS3 > 57 and high death risk; while piperacillin-tazobactam was prescribed to G2-patients as recommended in hospital for ICU septic patients related to low death risk based on SAPS3 score and TBSA 20-40%. Biomarkers C- reactive protein, leucocytes count, and creatinine clearance were monitored daily in ICU patients, and at TDM day blood sampling for simultaneous antimicrobials serum measurements purposes, based on drug prescription chosen for each septic major burn patient. It was found any difference between groups related to biomarkers at TDM, once cytokines storm happens at the earlier stage of septic shock with C-reactive protein and white blood cells increased.

Description of demographic and admission data; laboratorial data at TDM including ICU routine of biomarkers registered and coverage evaluated by PK/PD approach to investigate effectiveness during treatment with meropenem or piperacillin-tazobactam based on serum levels. Microbiology of isolated Gram-negative strains was described according CSLI database. All data for burn patients of both groups were detailed, Table 2.

Coverage was evaluated by the target attainment 100%$f\Delta T$ > MIC during the antimicrobial therapy done by 3hrs. infusion in major burns receiving meropenem (G1-patients n = 12) or piperacillin-tazobactam (G2-patients, n = 16). It was shown clinical cure by the piperacillin-tazobactam with eradication of all Gram-negative strains isolated up to MIC 16 mg/mL; in addition, clinical cure by the meropenem occurred by eradicating susceptible pathogens up to MIC 2 mg/L, as well extended to two isolates of P. aeruginosa and K. pneumoniae, both of intermediate susceptibility MIC 4 mg/L. Clinical outcome was investigated at the first septic shock that occurred after ICU admission by thermal injury with cure of all patients after eradication of Gram-negative pathogens. It is important to highlight that just one death was registered for a patient, due to a high total body surface area (> 70%), SAPS3 68 with high death risk associated. The length of stay period required for a patient in ICU takes about 30 to 45 days undergoing intensive care. Then, each major burn patient to be discharged occurs after completing all the necessary surgical procedures including the treatment of several septic shocks that occur between surgeries for each major burn patient considered. However, everything will depend on the recovery capacity of each patient burned by the fire. On the other hand, the length of stay of a patient in the ICU is also mainly related to the total area burned and the depth of the burn.

Table 2. Demographic, laboratorial data, PK/PD approach for antimicrobial effectiveness done by extended 3hrs.-infusion

Parameters	Piperacillin-Tazobactam Dose 4.5g q6h	Meropenem Dose 1g q8h
Demographic data	G1 (n = 16)	G2 (n = 12)
Gender (19M/9F)	10M/6F	9M/3F
Age (yrs)	29 (25-32)	32(28-34)
Body weight (kg)	68 (64-73)	69(66-75)
Height (cm)	172 (166-174)	169 (165-177)
Admission data		
TBSA (%)	33(28-39)	53(42-62) *
SAPS3	42(39-44)	57(51-68) *
Death risk (%)	23(16-29)	52(45-59) *
Inhalation injury (23/28)	12/16	11/12
Thermal burn (28/28)	16/16	12/12
Mechanical ventilation (27/28)	15/16	12/12
Vasopressors (26/28)	14/16	12/12
Biomarkers at TDM (CSLI, reference)		
C- reactive protein (mg/L) increased	141(97-315)	192(137-297)
Leucocytes (*1000 cells/mm3) increased	16.9(14.4-23.6)	14.8(11.9-22.6)
Creatinine clearance (mL/min) increased	179(132-253)	192(177-229)
PK/PD approach-coverage		
Coverage (MIC) (28/28)	2 mg/L (16/16)	1 mg/L (12/12)
Coverage (MIC) (28/28)	4 mg/L (16/16)	2 mg/L (12/12)
Coverage (MIC) (28/28)	8 mg/L (16/16)	4 mg/L (12/12)
Coverage (MIC) (28/28)	16 mg/L (16/16)	8 mg/L (12/12)
Clinical outcome		
Clinical-Microbiological cure (28/28)	16/16	12/12
Negative cultures up to (days)	12-14	10-12
ICU period (days)	(28-45)	(32-49)
Hospitalization (total days)	(37-62)	35-55)
Survivals (27/28)	15/16	12/12
No-Survivals (1/28)	1/16	0/12
Gram-negative isolates		
Enterobacteriaceae (48/80)	25/42	23/38
Non-Enterobacteriaceae (32/80)	17/42	15/38
Total of isolates (80)	42	38

Abbreviations: G1: group 1; G2: group 2; M: male; F: female; TBSA: total burn surface area; SAPS3: Simplified acute physiologic score; TDM: therapeutic drug monitoring; PK/PD: pharmacokinetics/pharmacodynamics; MIC: minimum effective concentration, based on CSLI database; CSLI: Clinical Laboratory Standard Institute; ICU: Intensive Care Unit.

Statistics: data by medians(quartiles), (*) $p < 0.05$ significant difference. GraphPad Prism v.9.4.1 software.

Thus, other factors may influence the recovery of major burn patients as the comorbidities, immunosuppression, and the severity of septic shock dependent of intensity of the systemic inflammatory response that occurs during each septic shock. All these factors may also contribute for the prolongation of the patient's stay in the ICU.

Conclusion

Bioanalytical method using the UPLCMS/MS technique was successfully developed and validated, showing a run time lower than 5 min. per assay. Main advantage of UPLCMS/MS is the simultaneous measurement of multiple β-lactam antibiotics, high selectivity and specificity, simple extraction procedure, short analytical run, and the small serum volume required. It was shown a good correlation between both liquid chromatographic techniques. Another advantage of UPLCMS/MS technique should be its application in a routine of therapeutic drug monitoring for these two widely prescribed antibiotics considering patients undergoing therapy of septic shock in the Intensive Care Units of hospital, caused by nosocomial Gram-negative pathogens. The new bioanalytical method based on UPLCMS/MS technique was applied for meropenem and piperacillin serum measurements in septic burn patients by applying the strategy of the extended infusion to attain the therapeutic target recommended for drug effectiveness by PK/PD approach; it is an important tool for the combat of bacterial resistance, also to avoid multidrug resistance and bacterial emergency resulting in death increases in ICUs.

It is important to highlight that clinical cure occurred for all patients by prescription of both antimicrobial agents through 3hrs.-infusion of dose regimens recommended by eradication of Gram-negative pathogens isolated. The application of this tool has been shown to be effective in combating the development of bacterial resistance in this ICU of the tertiary public hospital.

Disclaimer

None.

References

Abdul-Aziz MH, Lipman J, Mouton JW, Hope WW, Roberts JA. Applying pharmacokinetic/pharmacodynamic principles in critically ill patients: optimizing efficacy and reducing resistance development. *Semin. Respir. Crit. Care Med.* 2015 Feb;36(1):136-53. Available in: http://doi.org/10.1055/s-0034-1398490.

Abdul-Aziz MH, Lipman J, Akova M, Bassetti M, De Waele JJ, Dimopoulos G, Dulhunty J, Kaukonen KM, Koulenti D, Martin C, Montravers P, Rello J, Rhodes A, Starr T, Wallis SC, Roberts JA; DALI Study Group. Is prolonged infusion of piperacillin/tazobactam and meropenem in critically ill patients associated with improved pharmacokinetic/pharmacodynamic and patient outcomes? An observation from the Defining Antibiotic Levels in Intensive care unit patients (DALI) cohort. *J. Antimicrob. Chemother.* 2016 Jan;71(1):196-207. Available in: http://doi.org/10.1093/jac/dkv288.

Abdulla A, Ewoldt TMJ, Hunfeld NGM, Muller AE, Rietdijk WJR, Polinder S, van Gelder T, Endeman H, Koch BCP. The effect of therapeutic drug monitoring of beta-lactam and fluoroquinolones on clinical outcome in critically ill patients: the DOLPHIN trial protocol of a multi-centre randomised controlled trial. *BMC Infect. Dis.* 2020 Jan 17;20(1):57. Available in: https://doi.org/10.1186/s12879-020-4781-x.

Barlam TF, Cosgrove SE, Abbo LM, MacDougall C, Schuetz AN, Septimus EJ, Srinivasan A, Dellit TH, Falck-Ytter YT, Fishman NO, Hamilton CW, Jenkins TC, Lipsett PA, Malani PN, May LS, Moran GJ, Neuhauser MM, Newland JG, Ohl CA, Samore MH, Seo SK, Trivedi KK. Implementing an Antibiotic Stewardship Program: Guidelines by the Infectious Diseases Society of America and the Society for Healthcare Epidemiology of America. *Clin. Infect. Dis.* 2016 May 15;62(10):e51-77. Available in: http://dx.doi.org/10.1093/cid/ciw118.

Cain SE, Kohn J, Bookstaver PB, Albrecht H, Al-Hasan MN. Stratification of the impact of inappropriate empirical antimicrobial therapy for Gram-negative bloodstream infections by predicted prognosis. *Antimicrob. Agents Chemother.* 2015 Jan;59(1):245-50. Available in: https://doi.org/10.1128/aac.03935-14.

Camargo TV, Silva Jr., EM, Silva Jr., JM, Gomez, DS., Santos, SCJ. PK/PD approach to evaluate Meropenem effectiveness in critically ill burn adolescents versus young adults undergoing therapy of septic shock. *Pharm. Pharmacol. Int. J.* v.10(3): p.79–85, 2022. Available in: doi: 10.15406/ppij.2022.10.00368.

Carlier M, Stove V, Wallis SC, De Waele JJ, Verstraete AG, Lipman J, Roberts JA. Assays for Therapeutic Drug Monitoring of β-Lactam Antibiotics: A Structured Review. *International Journal of Antimicrobial Agents*. v. 46(4) p. 367–375. October 2015. Available in: http://doi.org/10.1016/j.ijantimicag.2015.06.016.

Caro YS, Cámara MS, De Zan MM. A review of bioanalytical methods for therapeutic drug monitoring of beta-lactam antibiotics in critically ill patients: Evaluation of the approaches used to develop and validate quality attributes. *Talanta* 210 (2020) 120619. Available in: <hppts://doi.org/10.1016/j.talanta.2019.120619>.

Casals G, Hernández C, Hidalgo S, Morales B, López-Púa Y, Castro P, Fortuna V, Martínez JA, Brunet M. Development, and validation of a UHPLC diode array detector method for meropenem quantification in human plasma. *Clin. Biochem.* v.47, p.223–227, 2014. Available in: https://doi.org/10.1016/j.clinbiochem.2014.08.002.

Chung EK, Cheatham SC, Fleming MR, Healy DP, Shea KM, Kays MB. Population Pharmacokinetics and Pharmacodynamics of Piperacillin and Tazobactam Administered by Prolonged Infusion in Obese and Nonobese Patients. *The Journal of Clinical. Pharmacology.* v. 55(8), p.899–908. May 2015. Available in: <http://doi.org/ 10.1002/jcph.505>.

CLSI (Clinical and Laboratory Standards Institute). 2007. *Mass spectrometry in the clinical laboratory: general principle and guidance; Approved guideline document C50-A;* First edition (ISBN 1-56238-648-4), Wayne PA: CLSI. http://www.clsi.org.

CLSI (Clinical and Laboratory Standards Institute). 2009. *Approved Guideline EP-6; Evaluation of linearity of quantitative methods,* Second Edition, Wayne PA: CLSI. http://www.clsi.org.

CLSI (Clinical and Laboratory Standards Institute). 2014. *User verification for Precision and estimation of Bias; Approved Guideline, CLSI document EP-15 A3, First edition,* Wayne PA: CLSI. http:// www.clsi.org.

CLSI (Clinical and Laboratory Standards Institute). 2014. *Liquid Chromatography-Mass Spectrometry Methods; Approved guideline EP62-A; First edition,* Wayne PA: CLSI. http:// www.clsi.org.

D'Cunha R, Bach T, Young BA, Li P, Nalbant D, Zhang J, Winokur P, An G. Quantification of Cefepime, Meropenem, Piperacillin, and Tazobactam in Human Plasma Using a Sensitive and Robust Liquid Chromatography-Tandem Mass Spectrometry Method, Part 1: Assay Development and Validation. *Antimicrob. Agents Chemother.* 2018 Aug 27;62(9):e00859-18. Available in: <http:// doi: 10.1128/ AAC.00859-18>.

Evans L, Rhodes A, Alhazzani W, Antonelli M, Coopersmith CM, French C, Machado FR, Mcintyre L, Ostermann M, Prescott HC, Schorr C, Simpson S, Wiersinga WJ, Alshamsi F, Angus DC, Arabi Y, Azevedo L, Beale R, Beilman G, Belley-Cote E, Burry L, Cecconi M, Centofanti J, Coz Yataco A, De Waele J, Dellinger RP, Doi K, Du B, Estenssoro E, Ferrer R, Gomersall C, Hodgson C, Møller MH, Iwashyna T, Jacob S, Kleinpell R, Klompas M, Koh Y, Kumar A, Kwizera A, Lobo S, Masur H, McGloughlin S, Mehta S, Mehta Y, Mer M, Nunnally M, Oczkowski S, Osborn T, Papathanassoglou E, Perner A, Puskarich M, Roberts J, Schweickert W, Seckel M, Sevransky J, Sprung CL, Welte T, Zimmerman J, Levy M. Surviving Sepsis Campaign: International Guidelines for Management of Sepsis and Septic Shock 2021, *Critical Care Medicine:* v.49(11) p. 1063-1143. November 2021 Available in: http://doi.org/10.1097/CCM.0000000000005337.

FDA (US Food and Drug Administration) *Guidance for Industry–Bioanalytical Method Validation* (BMV). 2020. FDA. https://www.fda.gov/regulatory-information/search-fda-guidance-documents/bioanalytical-method-validation-guidance-industry.

Fournier A, Eggimann P, Pagani JL, Revelly JP, Decosterd LA, Marchetti O, Pannatier A, Voirol P, Que YA. Impact of the introduction of real-time therapeutic drug monitoring on empirical doses of carbapenems in critically ill burn patient. *Burns* v.41, 2015. Available in: <hppt://dx.doi.org/10.1016;j.burns.2025.01.001>.

Global Sepsis Alliance. 4th World Sepsis Congress on April 25 and 26, 2023. *One Global Health Treat: Sepsis, Pandemics, and Antimicrobial Resistance.* Available in: https://www.worldsepsiscongress.org. global-sepsis-alliance.org. Available in: https://www.global-sepsis-alliance.org.

Global Sepsis Alliance. *Sepsis and Covid-19/Coronavirus/SARS-COV-2,* 2020. Available in: https://www.global-sepsis-alliance.org/covid19.

Goncalves-Pereira J, Silva NE, Mateus A, Pinho C, Povoa P. Assessment of pharmacokinetic changes of meropenem during therapy in septic critically ill patients. *BMC Pharmacol. Toxicol.* v.15:21, April 14, 2014. Available in: https:// doi:10.1186/ 2050-6511-15-21. http://www.biomedcentral.com/2050-6511/15/21

Karim, Najmul & Kabir, Alamgir & Islam, Md Manirul & Biswas, Akhil & Wasim, Mohammed & Alam, Mahbubul & Chowdhury, Nobendu & Islam, Mohammed & Tabassum, Tamanna & Hasan, Md. (2022). Use of cefepime, meropenem, or piperacillin/tazobactam as initial treatment for febrile neutropenia in patients with hematological malignancy — a real-life experience. *The Egyptian Journal of Internal Medicine.* 34. 10.1186/s43162-022-00154-4.Available in: https://doi.org/10.1186/ s43162-022-00154-4.

Lunn, G. 2005. *Methods for Recently Approved Pharmaceuticals,* first edition. Edit by John Wiley & Sons., New York, ISBN 9780471669418. Available in: http://www.doi:10. 1002/0471711683.

Messiano CG, Morales Junior R, Pereira GO, Silva Junior EMD, Gomez DS, Santos SRCJ. Therapeutic Target Attainment of 3-Hour Extended Infusion of Meropenem in Patients with Septic Burns. *Clin. Ther.* S0149-2918(22):00077–7. February 13, 2022. Available in: https://doi.org/10.1016/j.clinthera.2022.02.007.

Monti G, Bradic N, Marzaroli M, Konkayev A, Fominskiy E, Kotani Y, Likhvantsev VV, Momesso E, Nogtev P, Lobreglio R, Redkin I, Toffoletto F, Bruni A, Baiardo Redaelli M, D'Andrea N, Paternoster G, Scandroglio AM, Gallicchio F, Ballestra M, Calabrò MG, Cotoia A, Perone R, Cuffaro R, Montrucchio G, Pota V, Ananiadou S, Lembo R, Musu M, Rauch S, Galbiati C, Pinelli F, Pasin L, Guarracino F, Santarpino G, Agrò F E, Bove T, Corradi F, Forfori F, Longhini F, Cecconi M, Landoni G, Bellomo R, Zangrillo AContinuous vs Intermittent Meropenem Administration in Critically Ill Patients with Sepsis. The MERCY Randomized Clinical Trial. *JAMA*. doi: 10.1001/jama.2023.10598.

Oliveira CA and Mendes ME. 2011. Gestão da Fase Analítica do Laboratório, como assegurar a qualidade na prática *[Management of the Laboratory's Analytical Phase, how to ensure quality in practice],* volume I, online first edition, RJ: Edit by Controllab. http://www.controllab.com.br/pdf/gestaofaseanalitica vol1.pdf.

Rhodes A, Evans LE, Alhazzani W, Levy MM, Antonelli M, Ferrer R, Kumar A, Sevransky JE, Sprung CL, Nunnally ME, Rochwerg B, Rubenfeld GD, Angus DC, Annane D, Beale RJ, Bellinghan GJ, Bernard GR, Chiche JD, Coopersmith C, De Backer DP, French CJ, Fujishima S, Gerlach H, Hidalgo JL, Hollenberg SM, Jones AE, Karnad D R, Kleinpell RM, Koh Y, Lisboa TC, Machado FR, Marini JJ, Marshall JC, Mazuski JE, McIntyre LA, McLean AS, Mehta S, Moreno RP, Myburgh J, Navalesi P, Nishida O, Osborn TM, Perner A, Plunkett CM, Ranieri M, Schorr CA, Seckel MA, Seymour CW, Shieh L, Shukri KA, Simpson SQ, Singer M, Thompson BT, Townsend SR, Van der Poll T, Vincent JL, Wiersinga WJ, Zimmerman JL, Dellinger RP. Surviving Sepsis Campaign: International Guidelines for Management of Sepsis and Septic Shock: 2016. *Intensive Care Med.* 2017 Mar;43(3):304-377. doi: 10.1007/s00134-017-4683-6. Epub 2017 Jan 18.

Rudd KE, Johnson SC, Agesa KM, Shackelford KA, Tsoi D, Kievlan DR, Colombara DV, Ikuta KS, Kissoon N, Finfer S, Fleischmann-Struzek C, Machado FR, Reinhart KK, Rowan K, Seymour CW, Watson RS, West TE, Marinho F, Hay SI, Lozano R, Lopez AD, Angus DC, Murray CJL, Naghavi M. Global. Regional. and National Sepsis Incidence and Mortality. 1990–2017: Analysis for the Global Burden of Disease Study. *The Lancet.* v. 395(10219): p. 200–211. January 2020. Available in: http://dx.doi.org/10.1016/S0140-6736(19)32989-7.

Santos SRCJ, Camargo, TV, Macedo, VA, Pires, FR, Silva, JM, Gomez, DS. Vancomycin and meropenem serum monitoring for target attainment by PK/PD approach as an effective tool in the battle against nosocomial pathogens in septic pediatric burn patients. *Pharm. Pharmocol. Int. J.* v.10(3): p.87-95, May 24, 2022. Available in: https://doi: 10.15406/ppij.2022.10.00369.

Santos, SRCJ, Camargo, TV, Messiano, CG, Kupa, LVK, Souza, VK, Morales Jr, R, Pinto, DCS, Silva Jr, EM, Silva Jr, JM, Gomez, DS. Combating Bacterial Resistance to Antimicrobials in Severe Septic ICU Patients: Importance of Meropenem. Piperacillin Serum Monitoring as a Dose Adjustment and Duration of Infusion Strategies. *Pharm. Pharmacol. Int. J.* v.11(2): p.52–60, May 24, 2023a. Available in: https://doi: 10.15406/ppij.2023.11.00402.

Santos, SRCJ, de Camargo, TV, Messiano, CG, Kupa, LVK, Morales Jr, R, Souza, VK, Pires, FR, Pinto, DCS, Campos, EV, Silva Jr, EM, Silva Jr, JM, Gomez, DS. Combined vancomycin-meropenem therapy in pediatric major burns undergoing therapy of septic shock guided by cultures and pharmacokinetic-pharmacodynamics approach based on serum levels to combat bacterial resistance. *Pharm. Pharmacol. Int. J.* v.11(2):62–68. May 24, 2023b. doi: 10.15406/ppij.2023b.11.00403.

Sime FB, Roberts MS, Tiong IS, Gardner JH, Lehman S, Peake SL, Hahn U, Warner MS, Roberts JA. Can Therapeutic Drug Monitoring Optimize Exposure to Piperacillin in Febrile Neutropenic Patients with Haematological Malignancies? A Randomized Controlled Trial. *Journal of Antimicrobial Chemotherapy*. v. 70 (8): p. 2369–2375. May 2015. Available in: <http://dx.doi.org/10.1093/jac/dkv123>.

Taccone FS, Laterre PF, Dugernier T, Spapen H, Delattre I, Wittebole X, De Backer D, Layeux B, Wallemacq P, Vincent JL, Jacobs F. Insufficient β-Lactam Concentrations in the Early Phase of Severe Sepsis and Septic Shock. *Critical Care.* v.14(4): p. R126. 2010. Available in: <http://dx.doi.org/10.1186/cc9091>.

Udy AA, Lipman J, Jarrett P, Klein K, Wallis SC, Patel K, Kirkpatrick CM, Kruger PS, Paterson DL, Roberts MS, Roberts JA. Are Standard Doses of Piperacillin Sufficient for Critically Ill Patients with Augmented Creatinine Clearance? *Critical Care.* v.19(1): p. 28. 2015. Available in: <http://dx.doi.org/10.1186/s13054-015-0750-y>.

World Health Organization. 2023. *From emergency response to long-term covid-19 disease management: Sustaining gains made during the COVID-19 pandemic.* May 4th. doi: https://www.who.int/publications/i/item/WHO-WHE-SPP-2023.

Chapter 5

A Brief Review of Chromatographic Methods for the Quantification of Favipiravir

I. M. Nandeesha[1], MSc
G. Nagalakshmi[2], PhD
K. Umesha[3], PhD
B. C. Yallur[1], PhD
T. M. Veeresha[4], PhD
and D. H. Manjunatha[5,*], PhD

[1]Department of Chemistry, MS Ramaiah Institute of Technology, Bengaluru, Karnataka, India
[2]Department of Forensic Science, Soundarya Institute of Management and Science, Bengaluru, Karnataka, India
[3]PG Department of Chemistry, Vijaya College, Bangalore, Karnataka, India
[4]Department of Chemistry, Vijayanagar College, Hosapete, Vijayanagara, Karnataka, India
[5]Deparment of Studies in Chemistry, Shivagangotri, Davangere University, Davangere, Karnataka, India

Abstract

The chromatographic methods are crucial methods that aid in identifying the medication in both pure and formulation products, as well as in any other sample matrices. This technique involves many instruments like high-performance liquid chromatography (HPLC), ultra performance liquid chromatography (UPLC), gas chromatography (GC), liquid

* Corresponding Author's Email: manjunathdh@gmail.com; manjunathdh@davanagereuniversity.ac.in.

In: Chromatographic Methods and Research
Editor: Raven W. Obrien
ISBN: 979-8-89113-193-4

chromatography-mass spectroscopy (LC-MS/MS), high performance thin layer chromatography (HPTLC). These tools are renowned for determining the quality of pharmaceutical products and give the confidence to the end users. This chapter briefs about the determination of an antiviral drug, favipiravir (FVPR) in pure and pharmaceutical dosage forms and different biological matrix of samples, by employing various chromatographic techniques. FVPR is an antiviral medication that has shown promising results in treating a number of viruses that can be fatal, including the COVID-19 and Lassa as well as the Ebola viruses. For the FVPR analysis in matrices of pharmaceutical and biological sources, numerous analytical procedures have been developed and proven; however, new, sustainable and affordable methods must still be chosen as the protocol to boost its sensitivity, reliability, and repeatability. Chromatographic techniques are frequently used to assess FVPR in pharmaceutical and biological samples. Regulatory agencies like the World Health Organization (WHO), the United States Food and Drug Administration (USFDA), the Drug Control Government of India (DCGI), and others frequently approve these instruments. These techniques are used to ensure product quality in the pharmaceutical, food, and beverage industries. The present review chapter provides an overview of key analytical characteristics of these approaches.

Keywords: favipiravir, chromatographic methods, method development, validation

Introduction

In 1901, Russian botanist Mikhail Tswett developed adsorption chromatography while researching plant colours. By placing a concentrate of the leaves through a section of calcium carbonate, alumina, and sucrose, then eluting them with oil and ether/ethanol blends, he was able to extract different carotenoid and chlorophyll colors from leaves. From the Greek words chroma, which means "color," and graphos, which means "to write," he coined the name "chromatography" in 1906. His initial experiments were largely disregarded in the writing for a very long time. There are not many different chromatography types left today.

Chromatography is defined as "an actual strategy for detachment where the components to be isolated are dispersed between two phases, one of which is fixed (the stationary phase), while the other moves (the mobile phase) in clear course" by the World Association of the International Union of Pure and Applied Chemistry (IUPAC). Chromatographic techniques have significantly changed the capabilities of insightful science by allowing separation and investigation of extremely complex blends.

Gas chromatography (GC) and liquid chromatography (LC)are the two main types of chromatography. By separating the gas phase into a stationary phase and the vapour phase, gas chromatography separates vaporous compounds. Liquid chromatography includes the techniques such as size rejection (division is based on atomic size), particle trade (detachment is dependent on charge), while high-performance liquid chromatography (HPLC) technique is dependent on partitioning from a liquid phase. These techniques are also combined with thin layer chromatography (TLC), a planar type of LC, electrophoresis (where objectivity in an electrical field depends on the sign and size of solute charge), and particle chromatography (PC) (Christian, et al., 2013).

The Basic Principles of Chromatography Separation

In spite of the detail that various chromatography methods employ various retention mechanisms, they are all based on the dynamic distribution of an analyte between a stationary phase that is fixed and a mobile phase that is flowing. The preference of each phase for each analyte will be different.

A chromatographic column is used to sort components of a mixture. A small quantity of sample is placed at the top of the column before the stationary phase and solvent particles are added. Many people think that a "rate perspective" of chromatography is more rigorous than a "plate interpretation" of chromatography that is founded on equilibrium. According to this viewpoint, the division ratio is the most crucial element. It is merely the proportion of a duration of solute in the stationary phase to that in the mobile phase. In general, the formation of equilibrium between the stationary phase and the mobile phase is related to the retention of a compound by chromatography. True equilibrium may not be achieved, though, as chromatography is a dynamic process (Yadav and Singh, 2013).

Favipiravir

Favipiravir (FVPR) is an antiviral medication used to treat all three strains of influenza- A, B and C (Yadav and Singh, 2013). The IUPAC name of FVPR is 6-fluoro-3-hydroxy-2-pyrazine carboxamide, and its chemical formula is $C_5H_4FN_3O_2$ (Figure 1). This is a carboxamide derivative of pyrazine. The melting point is approximately 187^{0}C to 193^{0}C. It is slightly soluble in water but completely soluble in organic solvents including ethanol, DMSO and DMF. FVPR is a pro-drug that undergoes intracellular ribosylation and phosphorylation to form favipiravir ribofuranosyl-5'-triphosphate (favipiravir RTP), which inhibits viral RdRp (RNA-dependent RNA polymerase) and can bind, transcript, and replicate the viral genome while inhibiting viral RNA polymerase (Furuta, et al., 2017; Sissoko, 2016). Besides being important against influenza, the FVPR has been shown to have antiviral activity against alpha-, filo-, bunya-, arena-, flavi-noroviruses, and now COVID-19 (Aslam, 2020; De Clercq, 2019).

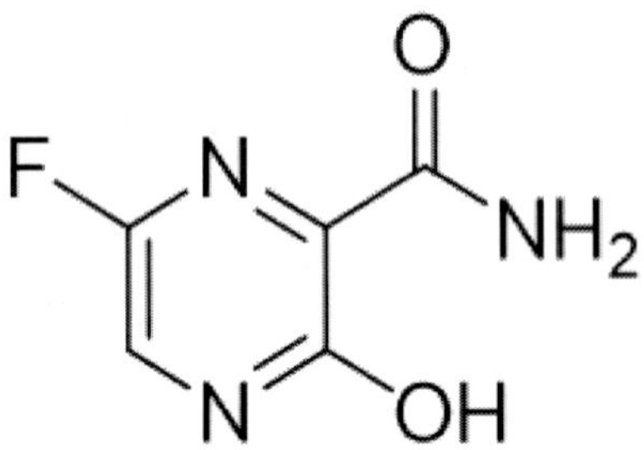

Figure 1. Chemical structure of favipiravir.

Mode of Action

The FVPR molecule is phosphoribosylated within the tissue to produce its active form, favipiravir-RTP. It works to fight viruses via the following mechanism (Furuta et al., 2017):

1. The RNA-dependent RNA-polymerase (RdRp) enzyme uses this molecule as a substrate, but the enzyme misinterprets it for a purine nucleotide, blocking its activity and stopping the production of viral proteins.
2. It is integrated into the viral RNA strand and stops the extension of the strand. The wide range of the efficacy of the drug is explained by this mechanism of action in combination with the persistence of the catalytic domain of the RdRp enzyme across many RNA viruses.
3. This virucidal drug has recently been found to trigger lethal mutagenesis in vitro during influenza virus infection. The physiological process of FVPR in shown in Figure 2.

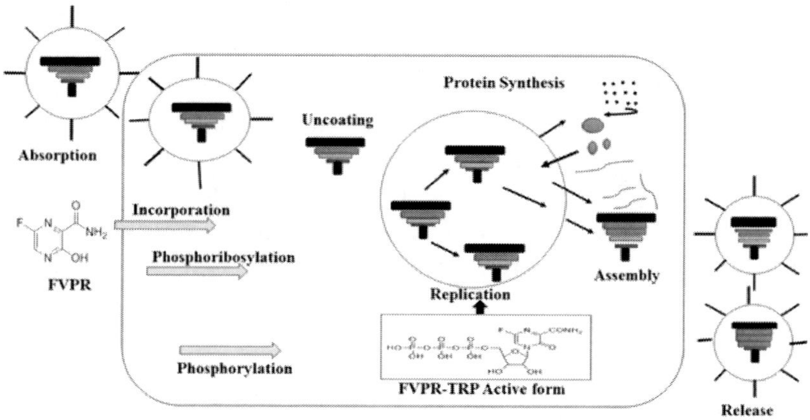

Figure 2. Physiological process of favipiravir (Furuta et al., 2017).

Physiological Properties and Stability of FVPR

FVPR prevent the viral genome from replicating. Antiviral activity of FVPR was reduced when purine bases or nucleosides were present, showing that these compounds compete with FVPR rather than pyrimidine nucleosides. Madin Darby Canine Kidney (MDCK) cells are useful for testing the influenza virus *in vitro*. FVPR was administered to MDCK cells, and then the cellular metabolites were examined using HPLC. There were three types of FVPR that were identified: FVPR ribofuranosyl-5′-triphosphate (RTP), FVPRribofuranose (R), and FVPR ribofuranosyl-5′-monophosphate (RMP). These findings imply that FVPR is activated after being absorbed into the cells. FVPR-RTP was chemically created, and its ability to stop the RNA polymerase of influenza virus activity was evaluated by incorporating P-GTP. In doses ranging from nanomolar to micromolar, FVPR-RTP reduced the viral RNA polymerase activity. At a concentration of 100 mol/L, neither FVPR nor FVPR-RMP had any effect on influenza RNA-dependent RNA polymerase (RdRp). Given that FVPR is intra-cellularlyphosphoribosylated to become the active form, FVPR-RTP, which inhibits viral replication by interacting with viral RNA polymerase. The findings from Yadav and Singh (2013) suggest that FVPR exerts its anti-viral activity as a pro-drug.It has not yet been completely determined how FVPR-RTP interacts with the RdRp molecule. The FVPR is expected to work by binding to conserved polymerase domains or by misincorporating in a nascent viral RNA to prevent the incorporation of nucleotides required for viral RNA replication and transcription. A recent study showed that FVPR lowered viral titer *in vitro* at low multiplicity of infection (0.0001 PFU/cell) or high multiplicity of infection (10 PFU/cell) and caused deadly mutagenesis during influenza virus infection. The nucleotide profiles of the NP gene analysed from multiple clones changed concurrently with an increase in the number of detectable G-A and C-T or C-U transition mutations, together with an increased mutation frequency, as a result of sequence analysis of different nucleoprotein (NP) clones (Yadav and Singh, 2013).

Pharmacology, Pharmacokinetics and Pharmacodynamics of FVPR

FVPR is a synthetic prodrug that was initially developed while testing the antiviral efficacy of chemical compounds of Toyoma Chemicals that were effective against the influenza virus. A/PR/8/34, a lead compound that was later known as T-1105, and its derivatives were discovered to have antiviral properties. The pyrazine moiety of T-1105 undergoes chemical alteration to produce favipiravir. In 2014, Japan licenced it for the treatment of newly emerging pandemic influenza infections (Yadav and Singh, 2013).

Prodrugs are used to give favipiravir. It has a high bioavailability (>94%), a diffident volume of distribution, and a 54% protein binding rate (10 - 20 L). After a single dose, it reaches C_{max} in less than two hours. T_{max} and half-life both grow after several doses. The hydroxylated version of favipiravir has a short half-life (2.5–5 h), which causes fast renal clearance. Aldehyde oxidase and xanthine oxidase, facilitate elimination to a lesser extent. The pharmacokinetics of favipiravir is dose and time-dependent. It inhibits one of the cytochrome components of P450 system but is not processed by it (CYP2C8). As a result, it should be provided with caution when combined with drugs that the CYP2C8 system metabolises (Yadav and Singh, 2013).

Analysis of FVPR by Various Chromatographic Techniques

Sample Treatment

The sample treatment plays an important role during determination of any drug by chromatographic method. The sample preparation will differ depending on the method used. According to the literature, following were the solution preparation procedures employed for the determination of FVPR.

Standard Stock Solution

Marzouk (2022) developed HPLC method for the determination of FVPR, wherein they accurately weighed 100 mg of pure medication and dissolved in 30 mL of deionized water and then transferred to a 100 mL volumetric flask. The volume was then increased to 100 mL using deionized water to create a stock solution containing 1 mg mL^{-1}. The resulting stock solution underwent sonication and a 0.45 μm filtering process. Before being injected into the system for analysis, the stock solution was further diluted with deionized water to achieve the necessary concentration of standard solutions (10-100 μgmL^{-1}).

In case of LC-MS/MS technique, Itigimath et al., (2021) have prepared standard stock solution of 1000 μgmL^{-1} concentration. 1.0 mL of the aforementioned stock solution was pipetted into a second 1000 mL standard volumetric flask and completed the task with the mobile phase. Similar to this, working standard solutions of various FVPR concentrations were prepared from least to greatest dilutions to examine the parameters of interest, such as linearity, accuracy, recovery, LOD, and LOQ of the proposed method. The concentration of the resulting working standard solution was 10 $ngmL^{-1}$.

Sample Solution- Pharmaceutical Formulation

In case of HPLC technique, a dry, clean mortar was used to transfer ten FVPR tablets, which were precisely weighed, and then grinded them into a fine powder. Next, a volumetric flask with a capacity of 250 mL was filled with tablet powder equivalent to 250 mg of FVPR. To evenly distribute the components, 100 mL of deionized water was added, and the flask was shaken on a rotary shaker for 10 minutes. After being sonicated for 30 minutes, the mixture was diluted to volume using deionized water to produce a solution that contained 1000 μg mL^{-1}, and then it was filtered through a 0.45 μm filter (Bulduk, 2021).

The test sample weights were collected based on the standard equivalent for the assay analysis, and the test sample weights were calculated using the formula given below (Itigimath et al., 2023):

$$sample\ weight = \frac{Standardweight\ x\ weightof\ 20\ tablets}{Label\ claim}$$

For the TLC method, the preparation of standard stock solutions of FVPR (1 mg/mL) involved weighing 100 mg of each substance into separate 100mL volumetric flasks, and then adding methanol to bring the volumes to the proper level. Ten FVPR tablets worth of their contents were weighed accurately and pulverised into a fine powder. Into a 25mL volumetric flask, placed a quantity equal to 25mg, followed by a 20minute sonication with the addition of 10 mL of methanol. The volume was made up to 25 mL mark with methanol, and a 0.45 μm membrane filter was used for filtering. To produce working solutions, specific quantities of each stock solution were diluted with methanol (Noureldeen, 2022).

Capillary micropipettes were employed for manual sample preparation during the conventional TLC analysis. Without the need for intermediary drying, sample sizes of 0.5 to 5 μL can be put as spots onto traditional layers. Lesser sample can be used on HPTLC plates (per spot, up to 1 μL is acceptable). After drying, the sample may be re-applied to the same area if more is required for detection or visualisation due to much diluted samples. Applying an automated instrumental sample using the spray-on technique is the sole option to carry out more challenging qualitative, quantitative, or preparative analyses or separations. Accurate automated positioning and volume dosing are essential for maximising separation power and repeatability in HPTLC method. Autosamplers/applicators are available at the high end, such as those from CAMAG that operate without an operator and apply sample as spots by spray-on approach, effectively using ink-jet printer-like technology. With the spray-on method, samples can be applied in bands or rectangles with quantities ranging from 0.5 μL to 50 μL. The best separation was provided by starting zones applied in the form of narrow bands. Rectangle application also enables accurate application of substantial quantities without compromising the layer. These rectangles are concentrated into small bands before chromatographic run using a solvent with high elution strength (Christian, et al., 2013).

Sample Solution-Blood Sample

A demonstration of the extraction of FVPR from plasma was made by Jain et al., in 2023. The following process was used: A human blood sample was centrifuged for 10 minutes at 5000 rpm to extract the plasma. The resultant sample was kept in storage at 4 0C until required and then gradually liquefied. To simulate drug-protein binding under physiological circumstances, biological samples including urine, blood, and plasma were applied with various doses of FVPR in the range of 0.01-10 μgmL-1, homogenised by vortex agitation for 5 min, and incubated at 37 0C for at least 30 min. These samples were utilised without any pre-treatment, such as deproteinization or filtering, for the purpose of validation (Jain et al., 2023).

Curley et al., (2021) have carried out the extraction of FVPR from rat plasma, human plasma and PBS (Phosphate buffer saline) as follows: 100 µl of standard, quality control (QC), blank or unknown sample was transferred to glass vials. These samples were diluted with 500 µL of MeOH (containing 10ng/mL of internal standard) and thoroughly vortexed. The samples were then centrifuged at 3500g for 5 minutes at 40C. 400 µL of supernatant fraction was transferred to a fresh glass vial and evaporated under a gentle stream of nitrogen. The resultant samples were reconstituted in 100µl of H_2O:MeOH (95:5) mixture. A 50µl of the sample was then transferred into 200µl chromatography vials and a 2µl of each sample was injected for analysis.

High Performance Liquid Chromatography (HPLC)

Amongst various chromatographic techniques employed for the determination of FVPR, HPLC acquires the most of it. About 60% of the research articles reports HPLC methods for its determination followed by LC-MS/MS, GC-MS/MS, GC, TLC and UPLC (Figure 3). The HPLC method reported by Nadendla and Abhinandana (2021) and Bulduk (2021) revealed quantification of FVPR in formulation by liquid chromatography method with PDA and UV detectors. Stability-indicating method for the determination of FVPR was developed by Marzouk et al., (2020) and Taskin (2022) used HPLC-DAD for the profiling dissolution study. Determination of FVPR by HPLC/UV in human plasma sample was reported by Abdallah, et al., (2022), Ali, et al., (2021) and Komarov (2022). They prepared the sample by liquid-liquid micro-extraction with methanol. QbD (Quality by Design) method has been developed by RP-HPLC for the determination of FVPR (Nishanth, et al., 2022; Amit, 2022; Aydinoglu et al., 2022; Srinivas, 2021; Taşkin, 2022; Shinde, 2023). This part of literature demonstrates how HPLC can be used to analyse samples of human blood plasma, formulation samples, and other matrices to determine an analyte.

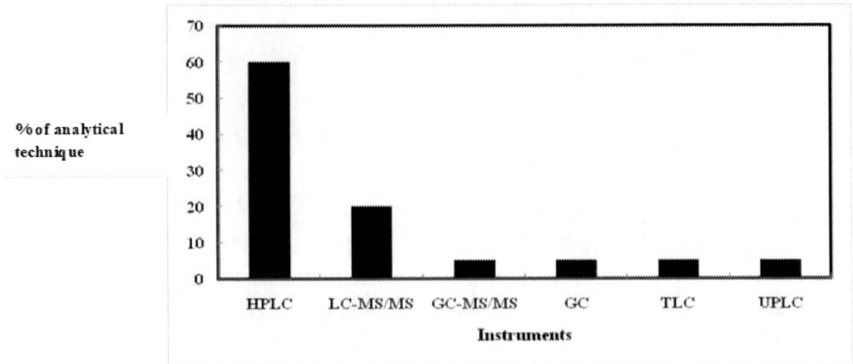

Figure 3. Graphical representation of reported chromatographic methods for the determination of favipiravir.

High-Performance Liquid Chromatography and Tandem Mass Spectrometry (HPLC-MS/MS)

The HPLC-MS/MS is quite sensitive than the other chromatography techniques. Few methods are available in the literature for the determination of FVPR by employing LC-MS/MS technique. The LC-MS/MS method developed and validated by Vemuri, (2022) and Abdallah et al., (2022) for the degraded impurities of FVPR in pure and solid dosage forms. An assay method has been developed and validated by Itigimath et al., (2022) using LC-MS/MS technique for the determination of FVPR in pure and dosage forms.

Gas Chromatography (GC)

The gas chromatography is mainly used for the estimation of residual solvents in drug substance and other materials like pesticides. This instrument can determine the organic solvents like methanol, ethanol, hexane, etc. The GC method reported by Vallakeerthi, et al., (2022) revealed the determination of residual solvents of FVPR. The biological and forensic samples of FVPR were determined by fabric phase sorptive extraction-gas chromatography-mass spectrometry (Jain et al., 2023).

High Performance Thin Layer Chromatography (HPTLC)

The HPTLC is the simple chromatography instrument, which helps to determine the drug substance instantly. The literature revealed that the determination of FVPR in blood plasma using the HPTLC instrument has been carried out by Noureldeen, (2022). This instrument is quite inexpensive than the other chromatography instruments like HPLC, UPLC, LC-MS/MS instruments.

Conclusion

The chromatographic techniques are separation methods. During olden days there were not much sophisticated instruments available for the estimation of drugs, drug impurities, food, beverages and residual solvents. Day by day improvisation found in the chromatographic techniques based on the regulatory guidelines and quality demand.

In the beginning, testing methods were done by TLC technique for qualitative estimation, later improved in the technology based on the same principle. Advanced equipments like the UPLC are created for quick and accurate results, while the HPLC devices are tailored for qualitative and quantitative estimation. On the other side, even though the LC-MS/MS instruments provide quick and accurate results for the qualitative and quantitative estimation, more money is necessary to pay for them. Instruments such as GC and GC-MS/MS were utilised to determine the residual solvents. The HPLC is more frequently employed for drug assessment in biological samples containing chemicals and other substances. While other instruments are also utilised, they are not used as frequently as HPLC, and HPLC is employed at a higher rate than the other instruments (Figure 3).

The FVPR drug is an anti Covid-19 drug. For its determination, many chromatographic techniques were used (Table 1) viz., HPLC, LC-MS/MS, GC, GC-MS/MS and TLC-densitometric. These techniques are advanced instruments and given the concordant results. But, the instruments were used based on the sample interest and availability. Some of the reported articles are revealed the assay, some determined the impurities, the GC instrument can be used to assess the residual solvents.

Table 1. Reported chromatographic methods of for the determination of favipiravir

Sl. No.	Chromato-graphic technique	Stationary phase	Mobile phase	Method	Linearity range, LOD and LOQ	Matrix	Reference
01	HPLC	Shim-Pack GIST C18 (250X 4.6 mm, 5μm)	10mM potassium dihydrogen orthophosphate buffer (pH 4.0) and acetonitrile in the ratio of 90:10 v/v	Assay method	10-60 μg/ml, 0.18 μg/ml, and 0.53 μg/ml.	pharmaceutical formulations	Nadendla, and Abhinandana, 2021.
02	HPLC	C18, (250X 4.6 mm, 5μm) column	Assay	Assay method	10–100 μg/ml, 1.20 μg/ml and 3.60 μg/ml	pharmaceutical formulations	Bulduk İbrahim, 2021.
03	HPLC-DAD	Zorbax C18, (250X 4.6 mm, 5μm)	0.1% (w/v) heptane sulphonic acid sodium salt-methanol–acetonitrile (62:28:10, by volume)	Assay method	6.25–250.00 μg/ml	pharmaceutical formulations	Marzouk, et al., 2022.
04	HPLC/UV	C18, (250X 4.6 mm, 5μm)	500 μl of tetrahydrofuran as an extractant and 1400 mg of fructose as a phase-separating agent	Degradation method	Linearity: 25-80,000ng/ml LOD: Not reported LOQ: Not reported	Human plasma	Abdallah, et al., 2022.
05	RP-HPLC	C18 column (5μm, 100 × 4.6 mm)	acetonitrile and ammonium acetate buffer (pH 4), in the ratio of 20:80v/v	QbD (Quality by design) method	Linearity: 0.062 - 4 μg/ml LOD: Not reported LOQ: Not reported	Pure drug	Nishanth, et al., 2022.
06	RP-HPLC	Cosmosil C18 (250 mm × 4.6ID, Particle size: 5 micron)	Methanol: Water (75:25, v/v). pH of Mobile Phase: 3 (pH is adjusted with o-phosphoric acid)	Quantitative method	Linearity: 10–50 ppm LOD: 0.2236 LOQ: 0.6776	pharmaceutical formulations	Amit,2022.

Sl. No.	Chromato-graphic technique	Stationary phase	Mobile phase	Method	Linearity range, LOD and LOQ	Matrix	Reference
07	HPLC	ACE-5 C18 column (250 mm × 4.6 mm, 5 μm),	10 mM phosphate buffer (pH = 2.5): methanol (80:20, v/v)	Quantitative method	Linearity range, 0.5 and 100 μg/ml LOD: 0.02 μg/ml LOQ: 0.05 μg/ml	pharmaceutical formulations	Aydinoglu, et al., 2022.
08	RP-HPLC	ODS column of dimensions 250x4.6 mm, 5μ	acetonitrile and 0.1% orthophosphoric acid (70:30)	Stability indicating method	Linearity: 10-150 μg/ml range, LOD: 1.818 μg/ml LOQ:6.248μg/ml	Pharmaceutical dosage form	Srinivas, et al., 2021.
09	HPLC-DAD	ODS column of dimensions 250x4.6 mm, 5 micron	mobile phase of acetonitrile and 0.1% orthophosphoric acid (70:30)	Quantitative method	Linearity: - Not reported LOD: Not reported LOQ: Not reported	Pharmaceutical dosage form	Taşkin, 2022.
10	RP-HPLC	buffer pH 3.5: acetonitrile [90:10]	ODS-3V C18 150 mm x 4.6mm x 5μ	forced degradation stability-indicating studies	Linearity: 50-250 μg/ml LOD:2.186 μg/ml LOQ:6.626 μg/ml	Pharmaceutical dosage form	Ali, et al., 2021.
11	HPLC-UV	C18, 150 × 4.6 mm, 5 μm	0.1% formic acid in water with 0.08% aqueous ammonia (eluent A), 0.1% formic acid in acetonitrile with 0.08% aqueous ammonia (eluent B)	FVPR quantified in human Plasma	Linearity: 0.25–200 μg/ml LOD: Not reported LOQ: 25 μg/ml	Human plasma	Komarov, et al., 2022.

Table 1. (Continued)

Sl. No.	Chromato-graphic technique	Stationary phase	Mobile phase	Method	Linearity range, LOD and LOQ	Matrix	Reference
12	RP-HPLC	C18 (Sunfire), 5 µ.m, 4.6 mm x 250 mm	the mobile phase used in pump A was ammonium acetate buffer of pH-6.5 and methanol in pump B. (Gradient mobile phase)	Assay method for Pure and formulation of FVPR	Linearity:-10 to 50 µg mL^{-1} LOD: -3.5 μgmL^{-1} LOQ: -12 μgmL^{-1}	Pure and pharmaceutical dosage form	Itigimatha, et al., 2022.
13	RP-HPLC	C8, 250 × 4.6 mm i.d., 5 µm	methanol:acetonitrile (50:50, % v/v)	QbD (Quality by design) method for assay	Linearity:20-100 µg/ml LOD: Not reported LOQ: Not reported	Pure and tablet dosage forms	Shinde, et al., 2022.
14	LC-MS/MS	AQUS C18, 250× 4.6 mm, 5µm particle	KH2PO4 buffer (pH 2.5 ± 0.05) and acetonitrile in the ratio of 98:2 (v/v), and mobile phase B contained water and acetonitrile in the ratio of 50:50 (v/v).	degradation impurities: Identification and route of degradation mechanism	Linearity: Not reported LOD: Not reported LOQ: Not reported	A contained finished solid dosage form.	Vemuri, et al., 2022.
15	LC-MS/MS	Shim pack GISS, C18 100 mm× 2.1 mm, 1.9 µm	pump A. 10.0 mM ammonium acetate and in pump B. methanol	Assay method	Linearity:- 50-200 ngmL-1 LOD: 4.044 ng/ml LOQ: 12.253 ng/ml	Pure and tablet dosage form.	Itigimath, et al., 2022.
16	LC-MS/MS	Kinetex F5 column 2.1×100mm 2.6µm (Phenomenex CA, USA)	Mobile phases A and B consisted of H2O with 0.1% acetic acid and MeOH with 0.1% acetic acid	Quantification of FVPR in human and mouse blood plasma	Linearity:- 10.78–200 ng/ml LOD: Not reported LOQ: 0.78 ng/mL	Human blood plasma and mouse plasma.	Curley, et al., 2021.

Sl. No.	Chromato-graphic technique	Stationary phase	Mobile phase	Method	Linearity range, LOD and LOQ	Matrix	Reference
17	LC-MS/MS	Hypersil C18 column, 4.6 × 150 mm, 5 µ	Acetonitrile-5 mM potassium dihydrogen phosphate (pH 2.5) (50:50, v/v)	Degradation method	Linearity: 0.5-100 µg/ml LOD: Not reported LOQ: Not reported	Pharmaceutical dosage form.	Abdallah, et al., 2022.
18	TLC-densito-metric	normal phase TLC plates	Solvent mixture of ethyl acetate, methanol and ammonia (8:2:0.2 ratio)	Qualitative method for FVPR in human blood	Linearity: 0.08 – 5.00 µg/band LOD: 0.02 µg/band LOQ: 0.07 µg/band	Pharmaceutical formulations and human plasma.	Noureldeen et al., 2022.
19	GC-MS	30 m length ×0.25 mm internal diameter ×0.25 µm film thickness, stationary phase 5% phenyl and 95% dimethylpolysiloxane.	Helium used as carrier gas	Residual solvents of FVPR quantification	Linearity: 0.01–10 $\mu g mL^{-1}$ LOD: 0.001 LOQ: 0.0026	Favipiravir in biological and forensic samples.	Jain et al., 2023.
20	HS-GC-FID	DB-624column, 30m×0.53mm×1.4 µm coating thickness	Nitrogen used as carrier gas	Residual solvents of FVPR quantification	Linearity: Not reported LOD: Not reported LOQ: Not reported	Quantification of residual solvents in favipiravir.	Vallakeerthi, et al., 2022.

Figure 3 shows that the HPLC is ideal for the determination drug and their impurities. About 60 percent of the reported articles found on HPLC for the estimation of FVPR and followed by LC-MS/MS technique, UPLC, GC, GC-MS/MS. The majority of the monographs of the Indian, British, and American pharmacopoeias (Indian Pharmacopoeia, 2022; British Pharmacopoeia, 2021; and United States Pharmacopoeia, 2021) demonstrate HPLC techniques for the determination of FVPR. Each chromatographic method has the unique application for the estimation of chemical substance as well as biological samples like human blood plasma, animal blood samples and urinal samples. The published articles shows that the HPLC is preferable for the assay and their impurities in pure and pharmaceutical formulations and in case of UPLC method, drug samples can be quickly estimated with minimum quantity of sample. The biological samples and small size samples can be determined by LC-MS/MS instrument and additionally molecular weight of the content can also be estimated. GC instruments are useful for the estimation of residual solvents of principal drugs. The TLC-densitometric technique is useful for the qualitative estimation of drugs by identifying the bands against the standard bands of respective drugs.

Hence the chromatographic methods and instruments are employed in the pharmaceutical industry, drug testing laboratories, analytical research and development laboratories, food industries, beverage industries and clinical trial labs. Based on the requirements one can opt the different analytical instruments to assure the quality of the product and maintain healthy society.

Disclaimer

None

References

Abdallah, Inas A, Mohammed F, El-Behairy, Rasha M, Ahmed, and Marwa AAF. The anti-COVID-19 drug Favipiravir: Degradation, Method development, Validation, NMR/LC–MS characterization, and In-vitro safety evaluation. *Chemical Papers* (2022) 76(10):6415-6426.

Abdallah, Inas A, Sherin FH, AlaaB, and Fotouh RM. Menthol-assisted homogenous liquid-liquid microextraction for HPLC/UV determination of favipiravir as an antiviral for COVID-19 in human plasma. *Journal of Chromatography* (2022) 1189: 123087.

Ali, Sayyed NS, Lajporiya M, Manjra M, Patel S, Aejaz A. and Khan GJ. Analytical method development and validation and forced degradation stability-indicating studies of Favipiravir by RP-HPLC and UV in bulk and pharmaceutical dosage form. *Journal of Pharmaceutical Research International.* (2021) (33) 48:254-271.

Amit ST. Method development and validation of favipiravir by RP-HPLC, *Innovative Journal of Medical Sciences* (2022) 6:2.

Aslam, Saima, and Mandeep R. Mehra. COVID-19: yet another coronavirus challenge in transplantation. *The Journal of Heart and Lung Transplantation* (2020) 5:408-409.

Aydinoglu, Sabriye, and Metin B. Favipiravir determination in pharmaceutical formulation via HPLC chromatographic approach. *Iranian Journal of Chemistry and Chemical Engineering* (2022).

British pharmacopeia. *British pharmacopoeia commission* (2023) (5).

Bulduk İ. HPLC-UV method for quantification of favipiravir in pharmaceutical formulations. *Acta Chromatographica* (2022) (33) 3:209-215.

Christian, Gary D, Purnendu K, Dasgupta, and Kevin A, Schug. *Analytical chemistry.* John Wiley & Sons. (2013).

Curley, Paul, Megan N, Usman A, Lee T, Henry P, Helen Box, Rajith KR Rajoli et al., Development of a highly sensitive bioanalytical assay for the quantification of favipiravir. *bioRxiv* (2021).

De Clercq, Erik. New nucleoside analogues for the treatment of hemorrhagic fever virus infections. *Chemistry An Asian Journal* (2019) 14: 22, 3962-3968.

Furuta, Yousuke, Takashi K, and Takaaki Nakamura. Favipiravir (T-705), a broad spectrum inhibitor of viral RNA polymerase. *Proceedings of the Japan Academy* (2017) 201, 7: 449-463.

Indian pharmacopeia, *Indian pharmacopoeia commission* (2022) (1).

Itigimath N, Hadagali A, Basappa CY, Manjunatha DH. LC-MS/MS method development and validation for the determination of favipiravir in pure and tablet dosage forms. *Turkish journal of pharmacy and technology* (2022). doi: 10.4274/tjps.galenos.2022.75470.

Itigimatha N, Chadchan KS, Yallur BC, Hadagali MD. New analytical methods for the determination of new anti-viral drug favipiravir: a potential therapeutic drug against covid-19 virus, in bulk and dosage forms. *Pharmceutical Chemistry Journal.* (2023) 56(10):1419-1425.

Jain R, Jain B, Kabir A, Bajaj A, Ch R, and Sharma S. Fabric phase sorptive extraction-gas chromatography-mass spectrometry for the determination of favipiravir in biological and forensic samples. *Advances in Sample Preparation* (2023) 6:1.

Komarov GVV, Tatineni S, Chiriki DS, Varshini N, Hemanth VPR, and GurupadayyaBM. Multivariate optimization for determination of favipiravir, a SARS-CoV-2 molecule, by the reverse-phase liquid chromatographic method using a QbD approach. *Journal of Chromatographic Science* (2022) bmac041.

Komarov TN, Karnakova PK, Archakova OA, Shchelgacheva DS, Bagaeva NS, Shohin IE, Zaslavskaya KY, and Bely PA. Development and validation of HPLC-UV method for the determination of favipiravir in human plasma. *Drug development & registration* (2022) 11(3), 220-229.

Marzouk, Hoda M, Mamdouh R, Rezk, Amira S, Gouda, and Ahmed M, Abdel-Megied. A novel stability-indicating HPLC-DAD method for determination of favipiravir, a potential antiviral drug for COVID-19 treatment; application to degradation kinetic studies and in-vitro dissolution profiling. *Microchemical Journal* (2022)172: 106917.

Nadendla, R, and Abhinandana P.A Validated high Performance Liquid Chromatographic Method for the Quantification of Favipiravir by PDA Detector. *International Journal of Life Sciences and Pharmaceutical Research* (2021) 11(2), 181-188.

Nishanth GVV, Tatineni S, Chiriki DS, Varshini N, Hemanth V, and Gurupadayya BM. Multivariate optimization for determination of favipiravir, a SARS-CoV-2 molecule, by the reverse-phase liquid chromatographic method using a QbD approach. *Journal of Chromatographic Science* (2021)041.

Noureldeen, Deena AM, John MB, Adel S. Lashien AF, Abdel H, and Tamer ZA. Novel environment friendly TLC-densitometric method for the determination of anti-coronavirus drugs remdesivir and favipiravir Green assessment with application to pharmaceutical formulations and human plasma. *Microchemical Journal* (2022) (174) 107101.

Shinde, Jyoti, Pankaj KS, and Jaya S. Development and validation of RP-HPLC method for estimation of favipiravir API and it's tablet dosage form using quality by design approach. *Journal of Pharmaceutical Negative Results* (2023)4115-4126.

Sissoko, Daouda, Cedric L, Elin F, Abdoul BM, Abdoul HB, Sylvain Baize andAlseny MC, et al., 2016. Experimental treatment with favipiravir for Ebola virus disease (the JIKI Trial): a historically controlled, single-arm proof-of-concept trial in Guinea. *PLoS medicine* (2016) 13:3.

Srinivas LB, Neelu J. Stability Indicative and Cost Effective Analytical Method Development and Validation of Favipiravir and Peramivir in bulk and Pharmaceutical dosage form by Using RP-HPLC. *International Journal of Applied Pharmaceutics* (2021) (13) 4.

Taşkina D. Development and Validation of a Rapid HPLC-DAD method for determination of favipiravir in pharmaceutical formulation. *Clinical and Experimental Health Sciences* (2022) (12) 3: 648-652.

United States Pharmacopeia, *United States Pharmacopeia* – National Formulary Publication. 2023(4).

Vallakeerthi, Narmada, Ravindernath A, Revathy SM, and Muralidhar RP. HS-GC-FID method development and validation for quantification of residual solvents in favipiravir. *Journal of Pharmaceutical Negative Results* (2022) 3675-3685.

Vemuri, Divya K, Rambabu G, Naresh K, Ravindra M, and Naresh KK, 2022. Favipiravir (SARSCoV2) degradation impurities: identification and route of degradation mechanism in the finished solid dosage form using LC/LC–MS method. *Biomedical Chromatography* (2022) (36) 6:e5363.

Yadav, Lal DS. Organic spectroscopy. *Springer Science & Business Media* (2023).

Biographical Sketch

D. H. Manjunatha

Affiliation: Davangere University, Karnataka, India

Education: MSc, PhD

Business Address: Assistant Professor, Department of Studies in Chemistry, Davangere University, Karnataka, India

Research and Professional Experience:

Manjunatha DH obtained his PhD in Chemistry from Karnatak University, Dharwad, INDIA in 2009 and is currently the assistant professor of chemistry since 2019 in Davangere University, INDIA. He has authored TWO textbooks, 3 book chapters and published several research papers in peer reviewed international journals. Currently his h-index is 16 and i10 index is 20. He completed two postdocs, one year each at Pohang University of Science and Technology (POSTECH), South Korea, during 2012-13 and Universiti Teknologi Malaysia (UTM), Malaysia, during 2012-2014.

Professional Appointments:

- Research Associate, Apotex Pharmachem India Pvt Ltd, Bangalore (a pharmaceutical company) during 2008-2011.
- Postdoctoral Fellow, Pohang University of Science and Technology (POSTECH), South Korea, during 2012-2013.
- Postdoctoral Fellow, Universiti Teknologi Malaysia (UTM), Malaysia during 2013-2014.
- Assistant Professor, Department of Chemistry, MS Ramaiah Institute of Technology, Bangalore, during 2016-2019.
- Assistant Professor, Department of Studies in Chemistry, Davangere University, Davangere, INDIA, since 2019.

Honors:

Indian National Science Academy (INSA) visiting Scientist award-2021 FY 2021-22, to visit Indian Institute of Science, Bangalore, INDIA.

Publications from the Last 3 Years:

1) New analytical methods for the determination of new anti-viral drug, favipiravir: A potential therapeutic drug against COVID-19 virus, in bulk and dosage forms. Nandeesha Itigimatha, Kailash S. Chadchan, Basappa C. Yallur, Manjunatha D. Hadagali. *Pharmaceutical Chemistry Journal* (Springer) (ISSN:1573-9031; 0091-150X)2023, 56(10), 1419-1425. doi 10.1007/s11094-023-02807-2,IF: 0.837.
2) Monitoring the interactions between bovine serum albumin and ZnO/Ag nanoparticles by spectroscopic techniques.Abhishek Udnoor, Manjunath Lokolkar, Basappa C. Yallur, Raju Kale, Muttanagoud N. Kalasad, Umesha Katrahalli & Devagondanahalli H. Manjunatha. *Journal of Biomolecular Structure and Dynamics* (ISSN:0739-1102;1538-0254)2023, 41(1), 352-365. doi: 10.1080/07391102.2021.2006788; IF: 4.15.
3) Molecular spectroscopic and docking analysis of the interaction of fluorescent thiadicarbocyanine dye with biomolecule bovine serum albumin. Umesha Katrahalli, Govindaswamy Shanker, Debnath Pal, Manjunatha Devagondanahalli Hadagali. *Journal of Biomolecular Structure and Dynamics* (ISSN:0739-1102;1538-0254). 2022, DOI: 10.1080/07391102.2022.2158135; IF: 4.15.
4) Liquid Chromatographic Method Development and Validation for the Simultaneous Determination of Phenylephrine Hydrochloride, Paracetamol, Caffeine and Diphenhydramine Hydrochloride in Pure and Formulations. Nandeesha IM, Basappa C. Yallur, Manjunatha D. Hadagali. *Research J. Pharm. and Tech* (ISSN: 0974-3618; 0974-360X)2022; 15(9):3993-3998. doi: 10.52711/0974-360X.2022.00669;
5) Synthesis of Ag nanoparticles capped by thiol molecules. Maheshkumar, V. S. Patil, H. J. Amith Yadav, D. H. Manjunath, S. V. Halse, M. N. Kalasad. *Materials Today: Proceedings.*(ISSN: 2214-7853)2022 (Available online) doi: https://doi.org/10.1016/j.matpr.2022.06.289.
6) Simple and Sensitive RP-HPLC and UV Spectroscopic Methods for the Determination of Remogliflozin Etabonate in Pure and Pharmaceutical Formulations. Nandeesha Itigimatha, Kailash S. Chadchan, Basappa C. Yallur, Manjunatha D. Hadagali. *Turkish Journal of Pharmaceutical Sciences* (ISSN: 1304-530X) 2022, 19(2), 213-219doi:10.4274/tjps.galenos. 2021.55381; IF: 0.95.
7) Preparation of colloidal Ag nanoparticles. Santosh R. Mannopantar, H. H. Bendigeri, V. K. Kulkarni, V. S. Patil, D. H. Manjunatha, M. N. *Kalasad.*

Materials Today: Proceedings. (ISSN: 2214-7853)2022, 60(2) 1156-1159, doi: https://doi.org/10.1016/j.matpr.2022.03.152.

8) Design, Synthesis, Anticancer Evaluation, and Molecular Docking Studies of Novel BenzoxazoleLinked 1,3,4-Oxadiazoles.
Sushmitha Bujji, Praveen Kumar E, Sree Kanth Sivan, Manjunatha D. H. and Subhashini N. J. P. *Anti-Cancer Agents in Medicinal Chemistry* (ISSN: 1875-5992; 1871-5206)2022, 22(5) 933-942 doi: 10.2174/1871520621666210706120203; IF: 2.668.

Chapter 6

Advances in Quantitative Liquid Chromatography-Mass Spectroscopy for Central Nervous System Melanocortin Hormone Quantification

Renée Onnainty[1], PhD,
Guadalupe Herrera[2], PhD,
and Gladys E. Granero[1,*], PhD

[1]Unidad de Investigaciones y Desarrollo en Tecnología Farmacéutica (UNITEFA) - ONICET, Departamento de Ciencias Farmacéuticas, Facultad de Ciencias Químicas, Universidad Nacional de Córdoba, Córdoba, Argentina
[2]IFEC-CONICET, Depto. Farmacología Otto Orsingher, Facultad de Ciencias Químicas, Universidad Nacional de Córdoba, Córdoba, Argentina

Abstract

The melanocortin system is related to the regulatory balance of various physiological functions including the eating behavior through the melanocortin pathway by stimulating receptors with several endogenous melanocortin agonists. These agonists include the mature hormone α-melanocyte–stimulating hormone (α-MSH), processed by the hypothalamic pro-opiomelanocortin (POMC) neurons, and are a signal for decreasing energy intake. Besides, a functional POMC circuit was also described in the hippocampus. In addition to its effects in the control of intake, α-MSH has potent anti-inflammatory and neuroprotective functions. This review outlines key considerations for the analysis of the

* Corresponding Author's Email: glagranero@unc.edu.ar.

In: Chromatographic Methods and Research
Editor: Raven W. Obrien
ISBN: 979-8-89113-193-4

hormone α-MSH in brain tissue by liquid chromatography tandem mass spectrometry (LC-MS/MS) method.

Keywords: literature review, liquid chromatography tandem mass spectrometry, α-melanocortin hormone, brain tissue

Introduction

The melanocortins (α, ß and γ-melanocyte-stimulating hormones; MSHs) have shown extra hormonal effects and because of their effects on brain functions and synthesis in the central nervous system, they are referred to as "neuropeptides" (Bertolini et al., 2009). These neuropeptides have proved anti-inflammatory and neuroprotective actions through their interactions with the melanocortin receptor 4 (MC4R) (Ramírez et al., 2019). Within the family of these endogenous peptides derived from proopiomelanocortin (POMC) neurons is the α-melanocyte stimulating hormone (α-MSH) (Catania et al., 2008). In the CNS, α-MSH is an endogenous MC4R agonist and has a physiological effect on feeding and energy homeostasis mediate, at least partly, by MC4R expressed by astrocytes, microglia and hypothalamus (Ramírez et al., 2019). On the other hand, inflammation and oxidative stress are linked in obesity and alterations of the melanocortin system contribute to food intake deregulation (Mountjoy et al., 2017), which is exacerbated by a high fat diet (HFD) (Sun et al., 2020).

Several researchers have shown by using different animal models that α-MSH has potent anti-inflammatory properties. Moreover, α-MSH is involved in the control of food intake and body weight (Catania et al., 2000).

In order to develop innovative therapies to treat the increasingly obese populations and diseases associated with imbalance in energy homeostasis, it is being explored the role of α-MSH in hypothalamic inflammation in rodent models of diet induced obesity (Kumar et al., 2014). With this goal in mind, α-MSH has been the focus of many research studies and, therefore, many of these studies involve the measurement of α-MSH levels in the brain.

Many scientists have focused on the use of antibody-based, electrochemical, bioluminescent, or other biological assays to characterize neuropeptides (Phetsanthad et al., 2023). However, researchers encountered some difficulties for measuring the amount of α-MSH in brain samples with these classical analytical methods, such as western bots and enzyme-linked immunosorbent assays, which have led them to look for new technologies that offer greater advantages in relation to the classic methodologies (Phetsanthad et al., 2023).

For example, radioimmunoassays (RIAs) have the disadvantage of their high cost and inability to simultaneously study multiple analytes, even though they are techniques highly sensitive and selectives (Li and Sweedler, 2008).

In contrast to classical techniques, mass spectrometry (MS) has begun to meet all the necessary requirements for researchers to fully study neuropeptides. MS measurement is based on measuring the mass to-charge ratio (m/z) of an analytic of interest. Within the MS, the high resolution, accurate mass (HRAM) instrumentation allows for identification at the single stage MS level. Moreover, masses can also be selected for tandem MS (MS/MS). Therefore, MS provides an attractive ability to examine the full complement of neuropeptides qualitatively and quantitatively. However, some key issues must be taken into account for the proper handling and separation of the samples to acquire quality data.

Although MS is a technique with high sensitivity, rapidity and chemical specificity, the mass spectral characterization of analytes of interest from tissue samples can be complicated due to the presence of interfering compounds in the biological matrix (Li et al., 2008). Therefore, proper handling of samples, from dissection of the animal to analysis to obtain reliable results is a key to obtaining relevant results, for which different strategies can be followed for processing biological samples. On the other hand, after the extraction of the analytes of interest from samples, the chemical complexity of the crude extracts requires a separation step prior to MS detection. For that, the liquid chromatography (HPLC) fractionation coupled to MS detection offers several unique advantages.

This chapter outlines key considerations for the analysis of the hormone α-MSH and its analog compounds in brain tissue by liquid chromatography tandem mass spectrometry (LC-MS/MS) method.

Liquid Chromatography Tandem Mass Spectrometry (LC-MS/MS) Method to Measure α-MSH and Its Analogous Compounds in the Brain

Among all the bioanalytical techniques, liquid chromatography coupled with tandem mass spectrometry (LC-MS/MS) is one of the most important bioanalytical methodologies to determine the concentrations of target analytes in both biological fluids and tissue samples (Li et al., 2014).

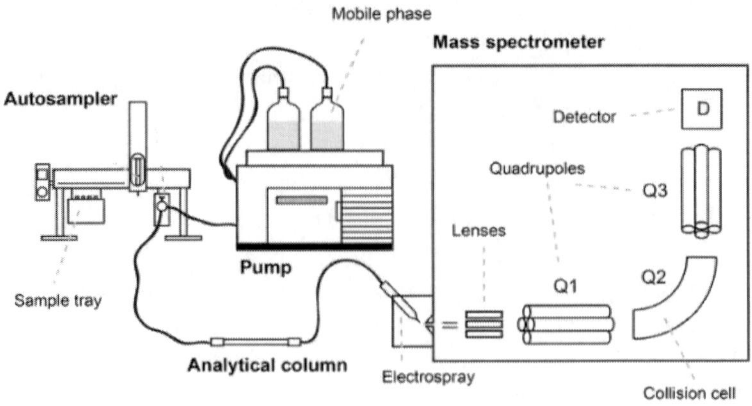

Figure 1. Schematic representation of LC-MS/MS.

In LC-MS/MS, first the analytes are separated from the biological matrix being processed by high performance liquid chromatography (HPLC) or ultrahigh performance liquid chromatography (UHPLC). After that, the eluting analytes are introduced into the mass spectrometer for their detection and quantification. Within the mass spectrometer, collision-induced dissociation occurs to generate different ionic fragments from the ionized precursor that are recorded, integrated and quantified.

Neuropeptides are implicated in a lot of functions in the brain and have pivotal roles in various neurological diseases; however, measuring their concentrations in the brain represents a great analytical challenge due to the low concentrations of these compounds found in the brain tissue. To address these drawbacks, LC-MS/MS is an emerging technique that allows the detection of these compounds in the pictogram order. Unlike immunoassays, LC-MS/MS allows the detection of specific sequences, which is important because there are many subtle variations of peptides (Zhou et al., 2015). Thus, the LC-MS/MS can provide high specificity due to additional structure information and high sensitivity.

Among the neuropeptides studied in the brain is the α-melanocyte-stimulating hormone (α-MSH) which exerts important functions including anti-inflammatory, antipyretic and immunoregulatory functions (Evest et al., 2003).

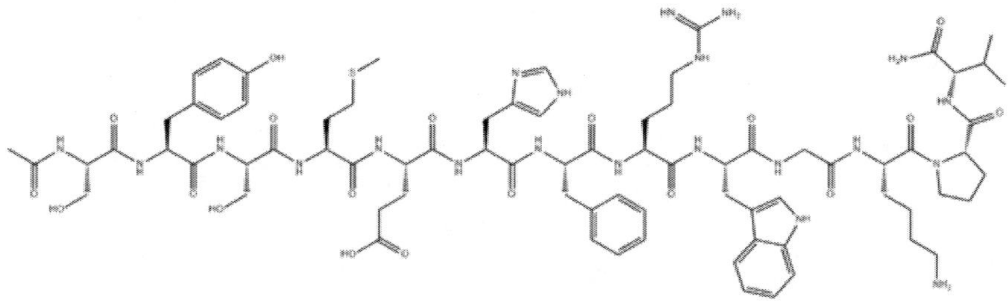

Figure 2. Chemical structures of α-MSH.

Some authors have also focused their researches on measuring the amounts of the synthetic analogue of the natural melanocortin peptide α-MSH, Melanotan-II (MT-II), which is known for its anorexic effects in rodents, which were attributed to the non-selective agonist of the melanocortin receptors (MCRs).

The direct measurement of MT-II in plasma and brain samples by LC/MS/MS was reported by Hatziieremia et al., (2007). These researchers wanted to find out if MT-II is able to penetrate the brain of rats. In this study, the chromatographic conditions were optimized by testing different column types, mobile phases, flow rates and elution gradients. Previously, Mock et al., (2002) also reported the determination of MT-II in rat plasma by LC/MS/MS. Trivedi et al., (2003) measured the MT-II levels by LC-MS/MS in specific regions of the hypothalamus following intravenous (*IV*) administration. These researchers intended to determine whether MT-II or its iodinated derivative (iodine-MT-II) is capable of penetrating the brain parenchyma to exert an anorectic effect.

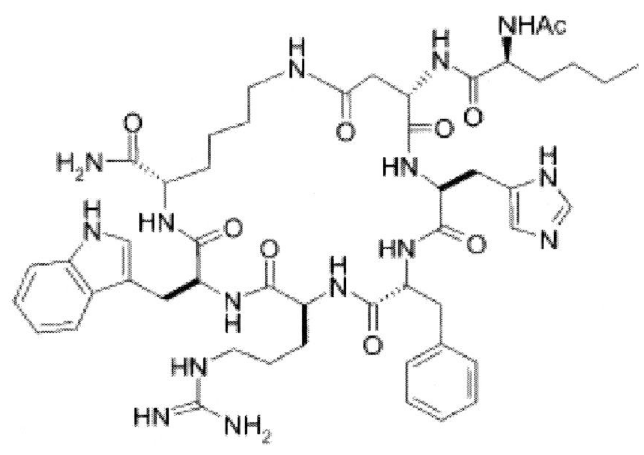

Figure 3. Chemical structure of MT-II.

Table 1. Summary of different chromatographic conditions for the determination of α-MSH and its derivatives by LC/MS/MS

Reference	Analyte	Sample	Column type	Mobile phase/Gradient	Flow rate	Temperature (°C)
Mock, 2002	MT-II	plasma		ACN/water (50:50, v/v) with 0.1% TFA	20 (μL/min)	
Trivedi, 2003	MT-II	Plasma brain	2.1 x 50 mm, 5 mm particle XDB-C18 Zorbax column	A: 5% ACN, 5 mM ammonium formate, 0.05% TFA; B: 90% ACN, 5 mM ammonium formate, 0.05% TFA/90% A, 10% B (0.1 to 0.8 min); 10% A, 90% B (0.8 to 2 min); 10% A, 90%B (2 to 3.5 min); 90% A to 10% B (3.5 to 4.5 min); 90% A, 10% B (4.5 to 10 min)	200 μL/min	
Hatziieremia, 2007	MT-II	Plasma brain	4.6 x 50 mm Symmetry 300 C4 (d_p = 5 mm) reversed-phase	A: 10% ACN, 90% water, 2 mM ammonium acetate, 0.1% FA and B: 90% ACN, 10% water, 2 mM ammonium acetate, 0.1% FA/0.1 to 1.0 min: 95% A:5% B, 1.0 to 5.0 min: 50% A:50% B, 5.0 to 8.0 min: 50% A:50% B, 8.0 to 10.0 min: 95% A:5% B, 10.0 to 15.0 min: 95% A:5% B.	0.6 mL/min	
Perroud, 2009	α-MSH deacetylated α-MSH acetylated α-MSH	brain	reverse phase C18 column, 55 μm, 70A	Gradient of ACN		
Kirwan, 2018	d-α-MSH α-MSH β-MSH	brain	2.1 x 50 mm 1.8 mm particle HSS T3	87.5% A (0.1% formic acid in water v/v) and 12.5% B (0.1% formic acid in ACN) 30 s before raising to 30% B over 2.5 min, flushed with 90% B for 1.1 min before returning to starting conditions	700 μL/min	60

Table 2. Summary of different mass spectroscopic conditions for the determination of α-MSH and its derivatives by LC/MS/MS

Reference	Analyte	Sample	detector	Ion mode	Voltage (V)	Temperature (°C)	Analysis mode
Mock, 2002	MT-II	plasma	triple quadrupole	positive	4350	400	Multiple reaction monitoring (MRM) mode
Trivedi, 2003	MT-II	Plasma brain	triple quadrupole	positive	5500	500	MRM
Hatziieremia, 2007	MT-II	Plasma brain	hybrid triple quadrupole/linear ion trap	positive	5500	550	MRM
Perroud, 2009	α-MSH deacetylated α-MSH acetylated α-MSH	brain	triple quadrupole				MRM
Kirwan, 2018	d-α-MSH α-MSH β-MSH	brain	triple quadrupole	positive	3000	600	MRM

Kirwan et al., (2018) with the aim of determining which POMC-derived peptides are present in human brain and their relative concentrations analyzed fresh post-mortem human tissue samples and human hypothalamic tissue and hypothalamic neurons derived from human pluripotent stem cells (hPSCs) by LC/MS/MS.

Perroud et al., (2009) investigated the effects *in vivo* of a serine protease, (PREP), on α-MSH by quantifying by mass spectrometry α-MSH and various forms of α-MSH, deacetylated α-MSH$_{1-13}$, acetylated α-MSH$_{1-13}$ and α-MSH$_{1-12}$.

The chromatographic conditions used in the previously cited reports are shown in Table 1. Whereas, the spectroscopic conditions used are displayed in Table 2.

Standard Internal (IS)

The use of internal standards (IS) is essential to the enhanced the performance of the method because the *IS* provides a measure of quality control for each analyte by compensating for changes in analyte, retention time, recovery, degradation, and changes in detector responses caused by coeluting contaminants (Phetsanthad et al., 2021).

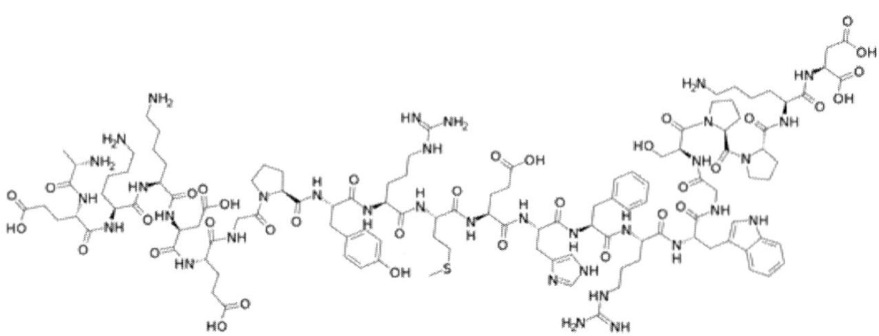

(A)

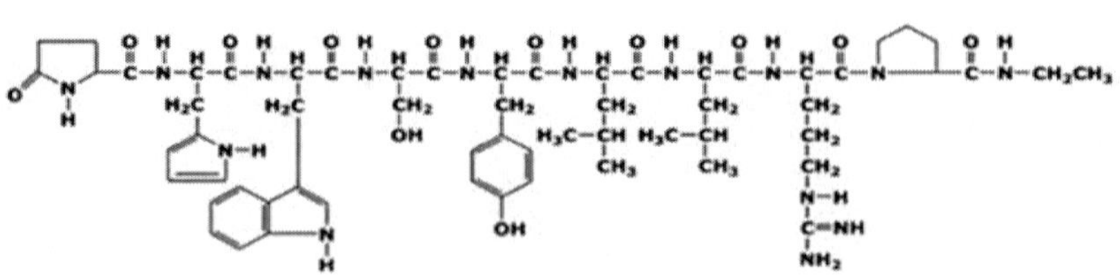

(B)

Figure 4. Chemical structures of internal standard (IS):
A) β-MSH and B) leuprolide.

The following compounds were used as internal standards (IS): leuprolide in the work reported by Hatziieremia et al., 2007, the iodo-MT-II in the report of Triveli et al., 2003 and the isotope-labeled α-MSH or β-MSH in the report of Kirwan et al., 2018.

Sample Preparation

Although, LC-MS/MS is a very sensitive and specific technique that allows the measurement of neuropeptide levels in the brain tissue, however, it is often frustrating the determination of these compounds due to matrix interferences. Therefore, to analyze brain tissue samples by LC-MS/MS, brain tissues must be adequately prepared with the objective of minimizing or suppressing the effect of the coeluting matrix components.

This prior sample preparation procedure is known as sample pretreatment or sample cleanup. This step can be considered as a pre-analytical separation process, consisting mainly of selective isolation of the analytes of interest from the matrix, minimization or elimination of matrix components and enrichment of the analytes.

One of the greatest challenges in brain samples is the high lipid component, which represents almost half of the dry weight of the brain tissue (Li et al., 2014). On the other side, brain tissue must be transformed into a liquid sample in order to be analyzed by LC-MS/MS, which is done with the help of techniques such as homogenization, microdialysis, ultrafiltration or solid state extraction (Li et al., 2014).

Sample Collection

For the purpose of measuring the amounts of neuropeptides in the brain tissue of model animals, samples of whole brains or certain brain regions are obtained after animal sacrifice. Because the brain is a solid tissue, tissue homogenization is commonly used.

Excised brain tissue should be frozen at -80°C or kept on ice to maintain a low temperature to keep brain tissue quality and analyte stability. To remove blood from the tissue, and thus increase the accuracy of analyte measurement, tissue samples are perfused with cold physiological saline or phosphate buffered saline (PBS). However, these procedure should be performed carefully so as not to remove the analytes of interest from the brain tissue samples.

Homogenization uses the mechanical force to disintegrate the brain tissue and disperse it in a suitable solvent, in order to obtain a suspension for future sample preparation. Also, a known amount of an analyte called internal standard (IS) is usually added to the homogenates obtained.

Homogenization of tissues can be carried out by different types of techniques, including grinding, rotary knife, bead beating and ultrasound. Because the brain tissue is a soft tissue after mechanical homogenization, brain tissue samples do not require further treatment. The result of the homogenization procedure is a homogeneous suspension which will be submitted to further sample preparation steps.

Extraction Procedures

There are different techniques for the pretreatment of fluid samples or biological tissues, among them are some of the following: Dilute-and-shoot protein precipitation (PPT), Liquid-liquid extraction (LLE) and Solid phase extraction (SPE) (Li, 2014).

In the published work of Hatziieremia et al., (2007), the extraction of MT-II from brain homogenates was performed by a procedure of precipitation with acetonitrile, centrifugation and subsequent evaporation of the supernatants.

Whereas, in the works reported by Trivedi et al., (2003) and Kirwan et al., (2018), the extraction of MT-II was performed by a solid-phase extraction method.

Mass Spectrum Ion Detection

In the mass spectrometer, the detection and quantification of the analytes can be made by using a multiple reaction monitoring (MRM), also referred to as selected reaction monitoring (SRM) function (Li et al., 2014).

Almost all reports cited herein measured the levels of α-MSH or its analogous compounds in brain samples by LC-MS/MS choosing a multiple reaction monitoring (MRM) method. MRM is the most common method for quantitation of analytes by LC-MS/MS, which requires a triple quadrupole MS that firstly targets the ion corresponding to the compound of interest with subsequent fragmentation of that target ion to produce a range of daughter ions.

In MRM, ions selected to make it through the first quadrupole and into the collision cell are referred to as the precursor, or parent, ions. These ions are fragmented in the collision cell. Certain fragment ions referred to as product, or daughter ions are selected to make it through the second quadrupole. After, one or more daughter ions can be selected for quantitation purposes. Only the selected parent ions and daughter ions corresponding to the mass of the molecule of interest are isolated within the mass spectrometer.

Thus, by ignoring all other ions that flow into the mass spectrometer the experiment gains sensitivity, whilst maintaining exquisite accuracy.

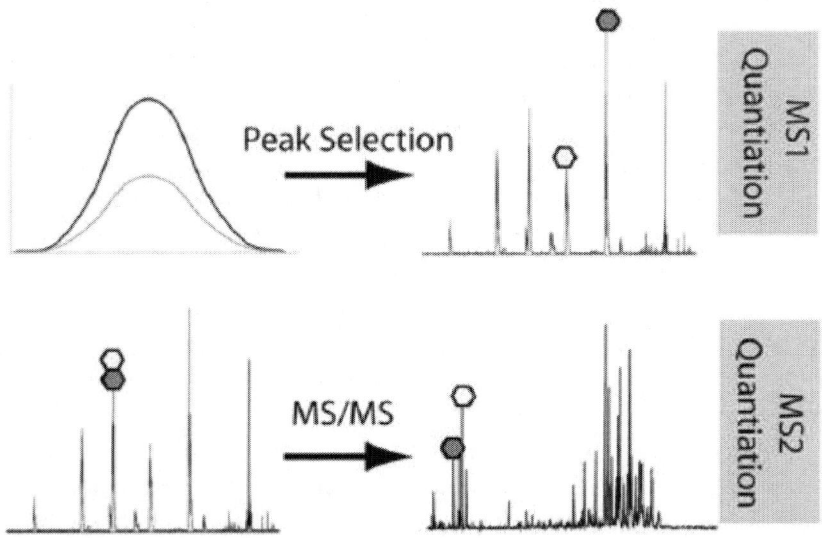

Figure 5. Detection and quantification of the analytes ions by LC-MS/MS.

The transition from precursor/parent ion to product/daughter ions is referred to as an ion transition. In the report of Hatziieremi et al., (2007), the following transitions were selected for quantification of MT-II in plasma and brain samples: m/z 513 → 110 and 513 → 128, respectively. Whereas, for the IS (leuprolide), the transition m/z 605 → 249 was monitored. Trivedi et al., (2003) also quantified MT-II by using the MRM mode, monitoring the transitions m/z 513.2 → 435.7, 513.2 → 413.0 and 513.2 → 128.1.

Kirwan et al., (2018) chose the MRM function to measure α-MSH and β-MSH by choosing the following transitions, m/z 555.98 and m/z 552.1, respectively. While choosing the m/z 554.6 ions for the internal standard.

Discussion

Research on the pleiotropic functions of melanocortins in the brain has been drawing significant attention from researchers over the last decade. In particular, the neuropeptide α-melanocyte–stimulating hormone (α-MSH) is an important anorexigenic neuromodulator in the hypothalamus but also various lines of evidence suggest that it is involved in the regulation of several central and peripheral physiological functions such as food intake and energy metabolism, inflammation and sexual dysfunction (Hatziieremi, 2007).

α-MSH is also involved in anti-inflammatory mechanisms through melanocortin 4 (MC4R) and possibly 1 (MC1R) receptors (Perroud, 2009). Because the inflammatory processes contribute widely to the development of neurodegenerative diseases, related investigations about the expression of many inflammatory mediators have had a very important boom in the last few years. Particularly, in the role of α-MSH and MC4R in inflammatory processes (Lasaga et al., 2008). Some researchers are exploring if α-MSH can modulate the long-term weight loss therapy responses in a dependent manner according to its concentration. It is therefore important to be able to quantify the concentrations of this neuropeptide in the brain regions involved in this function with the aim of finding therapeutic alternatives to obesity in clinical practice (Pelissari Kravchychyn et al., 2018).

There is a clear need for a sensitive technique capable of detecting very low levels of α-MSH in brain samples, since this neuropeptide, like its synthetic analogue MT-II, has a negligible degree of penetration into the brain (Hatziieremia et al., 2007). In this regard, the LC-MS/MS stands out in comparison to other techniques because it is sensitive and highly specific, providing information on the precise sequence of peptides in a single experiment (Kirwan et al., 2018), as opposed to ELISAs, which provide information about target peptides but cannot discriminate between peptides that share common epitopes but that might have functionally-relevant structural differences.

Following liquid chromatography (LC) separation, mass spectrometry (MS) is used to detect metabolic signal intensities that are proportional to analytes concentrations for quantitative assessments. In the MS, the analytes are subjected to high energy electron beams which results in ionization and fragmentation of parent molecules.

MS provides an alternative assay approach, relying on the discriminating power of mass analyzers to select a specific analyte and ion current measurements for quantitation. Most such assays use electrospray ionization followed by two stages of mass selection: a first stage (MS1) selecting the mass of the intact analyte (parent ion) and, after fragmentation of the parent by collision with gas atoms, a second stage (MS2) selecting a specific fragment of the parent, collectively generating a selected reaction monitoring (MRM) assay (Anderson et al., 2006).

LC-MS/MS analysis operating in MRM scan mode is a powerful technique due to its high sensitivity and selectivity. The MRM assay approach has been applied to the measurement of specific peptides in complex mixtures because it is able to quantify the targeted analytes at trace amounts in a variety of matrices (Matraszek-Zuchowska et al., 2016).

From a technical standpoint, the MRM mode consists of double mass filtration stages. In the first one, a specific precursor ion of interest is pre-selected in the first quadrupole (Q1) and the fragmentation is induced by collision excitation with a neutral gas in a pressurized collision cell (the second quadrupole Q2). In the second stage, fragments (product ions) generated in Q2 are analyzed in the third quadrupole (Q3).

Multiple reactions monitoring (MRM, also known as Selective Reaction Monitoring – SRM) makes LC-MS/MS a technique highly specific and sensitive, able to selectively quantify compounds within complex mixtures. Compared to radioimmunoassay (RIA), MRM is not restricted by the availability of good antibodies, which may represent a problem with proteins with low specificity. Moreover, RIA lacks the specificity required to distinguish between various biologically important forms of the targeted molecules (Perroud et al., 2009).

Another important aspect to consider in analyzing brain tissue samples by LC-MS/MS is the preparation of the sample to a form compatible with this technique and its choice will have a great influence on its sensitivity and robustness.

The process of preparing brain tissue samples for analysis by LC-MS/MS may involve various chemical or physical procedures to clean the sample and transform it into a liquid form (Li et al., 2014).

In the works cited in this manuscript, first, homogenates were obtained from brain tissue through a homogenization process. However, different methods were employed for the process of analyte extraction from the brain homogenates.

Hatziieremi et al., (2007) extracted the MT-II from brain homogenates with an acetonitrile (ACN) precipitation procedure. This method of protein precipitation (PPT) is commonly used due to its low cost and ease of implementation. The three stages of this technique are: addition of the precipitant, mixing and removal of the precipitated proteins.

In the report of Hatziieremi et al., (2007), ACN was used as the precipitating agent. The mixture was vortexes and solicited and lastly, the protein removal was carried out by centrifugation. Finally, the supernatant was evaporated to complete dryness and reconstituted into the mobile phase of LC.

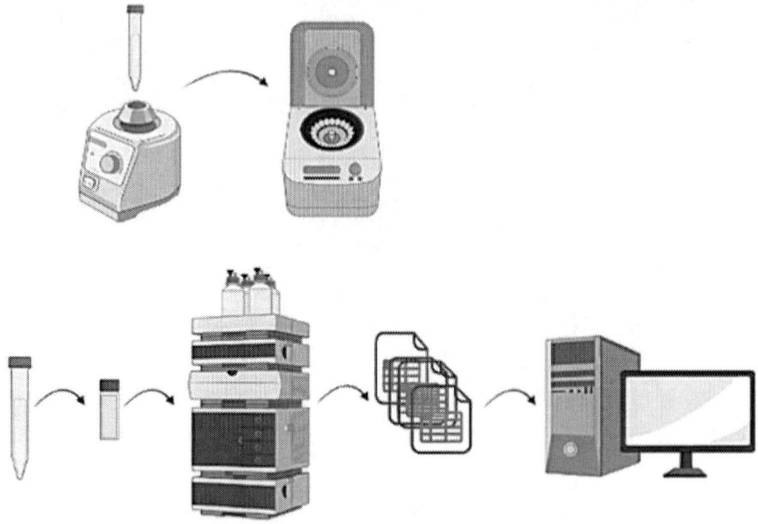

Figure 6. Diagrammatic representation of the process of protein precipitation (PPT).

Hatziieremi et al., (2007) evaluated different methods for the extraction of MT-II from brain homogenates. For solid phase extraction method (SPE), they used the Elution Oasis SPE plates commercially available and also, alternatively, they employed the protein precipitation method with an ACN precipitation protocol.

The yielded recovery of the analyte by using the SPE method was very low (40%). Whereas, by using ice-cold ACN followed by a filtration step, the recovery was improved significantly to 60%.

Kirwan et al., (2018) extracted the analytes from the brain homogenates by applying a SPE procedure. The basis of this technique is similar to that of liquid chromatography, which is based on the affinity between the analytes dissolved in a liquid (mobile phase) and the solid materials of the stationary phase. Depending on the affinity of the analytes of interest or unwanted impurities for the stationary phase, either the desired analyte or unwanted impurities in the sample will be retained in the stationary phase leading to separation of the components (Li et al., 2014).

Due to the large amount of insoluble cellular debris in brain homogenate samples, which can cause clogging of the stationary phase of the SPE, brain homogenates should be centrifuged beforehand. After that, the supernatant is forwarded to SPE. However, before passing the supernatants from the centrifuged brain homogenates, the SPE stationary phase must be prepared for activation by washing it with pure organic solvents, such as ACN or methanol

(MeOH). After activation of the SPE stationary phase, the samples are loaded to allow them to pass through the stationary phase SPE slowly by gravity.

Trivedi et al., (2003) extracted MT-II from brain homogenates using Waters Oasis HLB cartridges, which is a universal water wettable polymeric reversed-phase sorbent. In this report, the aqueous brain homogenates containing the analyte of interest (MT-II) and an SI were introduced into the cartridges, which were subsequently washed with water and eluted with 90% ACN and 0.1% TFA. Then, eluates were dried to dryness and reconstituted with the mobile phase of LC.

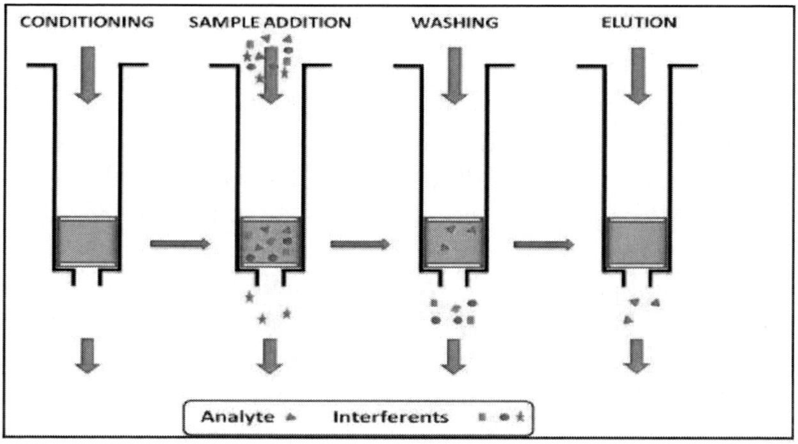

Figure 7. Schematic representation of the SPE process.

Figure 8. Waters Oasis HLB 12 cc Vac Cartridge.

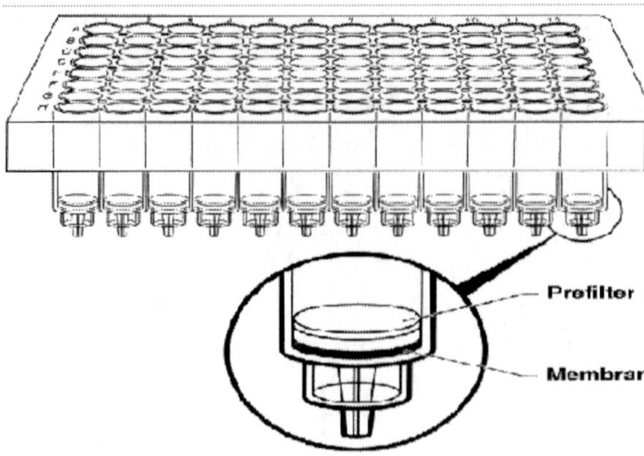

Figure 9. 96-Well SPE plate format of SPE.

Kirwan et al., (2018) used 96-Well SPE plate format (Plum et al., 1997).

In Kirwan's report, the SPE plates were first washed with 0.1% formic acid in dH_2O (v/v), followed by a wash with 1% acetic acid with 5% MeOH in dH_2O (v/v/v) to activate them. Samples were eluted with 60% MeOH in dH_2O with10% Acetic acid (v/v) and diluted further with of 0.1% formic acid in dH_2O (v/v) and stored at -80°C until loading onto LC columns for MS/MS.

Table 1 shows that different chromatographic conditions were used to separate the neuropeptides of interest.

There are several types of chromatography, each differing in the kind of stationary and mobile phase used. Kirwan et al., (2018) for instance carried out the chromatographic separations by using a silica-based bonded phase chromatographic column. Although these types of columns are compatible with an 100% aqueous mobile phase and are the first choice when developing separations for polar and non-polar compounds, these authors used as HPLC mobile phase for the separation a gradient between an acidic aqueous phase (0.1% formic acid) with ACN (0.1% formic acid).

The chromatographic columns whose stationary phase is silica are referred to as normal-phase chromatography, implying that the stationary phase is polar (hydrophilic) in nature and the mobile phase is nonpolar (hydrophobic) in nature; while the rest of the reports cited in Table 1 used chromatographic columns referred to as reversed-phase columns. In this case the scenario reverses, i.e., the stationary phase is nonpolar while the mobile phase is polar. On the other hand, ACN was used as the most lipophilic organic solvent in the composition of the mobile phase for HPLC separation.

Despite the fact that there are several types of chromatography, each differing in the kind of stationary and mobile phase they use, the underlying principle though remains the same, is to say that differential affinities of the various components of the analyte towards the stationary and mobile phases results in the differential separation of the components.

Conclusion

At the moment, neuropeptides are the focus of many recent research studies, which are leading the scientist to look for techniques that are more advantageous than those commonly used to quantify these hormones, since many of them are found in very low concentrations in the body's tissues, especially the brain tissue.

Among neuropeptides, the α-Melanocyte stimulating hormone (α-MSH) is known to be an inflammatory mediator with potent anti-inflammatory actions. Therefore, the α-MSH may be a very interesting therapeutic strategy in the treatment of CNS disorders with an inflammatory component. However, this neuropeptide must be able to cross the blood-brain barrier and reach therapeutic concentrations at target sites.

The synthetic analogue of **α**-MSH, the neuropeptide Melanotan-II (MT-II), a cyclic heptapeptide, is a potent non-selective melanocortinergic agonist and it elicits a pronounced inhibitory effect on food intake in rodents. Consequently, many investigations require the direct measurement of this neuropeptide in the brain tissue.

In order to explore the degree of penetration of the neuropeptides melanocortins into the brain, some researchers have required direct measurement of these compounds in brain tissue or in specific brain regions.

LC-MS/MS methods have comparable sensitivity to most ELISA techniques and greater specificity, since antibodies may recognize multiple similarities.

LC-MS/MS is a sensitive and reliable quantitation approach that in the future will greatly benefit our understanding of neuropeptides and neurochemical signaling.

Disclaimer

None.

References

Bertolini A., Tacchi R., Vergoni A. V. Brain effects of melanocortins. *Pharmacological Research* (2009):13-47.

Catania A. Neuroprotective actions of melanocortins: a therapeutic opportunity. *Trends in Neurosciences* (2008) 31(7):353-360.

Catania A., Airaghi L., Colombo G., Lipton J. M. a-Melanocyte-stimulating Hormone in Normal Human Physiology and Disease States. *TEM* (2000) 11 (8):304-308.

Eves P., Layton C., Wagner M., H. Kemp, M. Szabo, R. Morandini, G. Ghanem, J. C. García-Borrón, C. Jiménez-Cervantes, and S. Mac Neil. Anti-inflammatory and anti-invasive effects of melanocyte stimulatin hormone in human melanoma cells. *British Journal of Cancer* (2003) 89:2004-2015.

Hatziieremia S., Kostomitsopoulos N., Balafas V., Constantin Tamvakopoulos A liquid chromatographic/tandem mass spectroscopic method for quantification of the cyclic peptide melanotan-II. Plasma and brain tissue concentrations following administration in mice. *Rapid Communications in Mass Spectrometry* (2007) 21:2431-2438.

Kirwan P., Kay R. G., Brouwers B., Vicente Herranz-Pérez, Magdalena Jura, Pierre Larraufie, Julie Jerber, Jason Pembroke, Theresa Bartels, Anne White, Fiona M. Gribble, Frank Reimann, I. Sadaf Farooqi, Stephen O'Rahilly, Florian T. Merkle. Quantitative mass spectrometry for human melanocortin peptides *in vitro* and *in vivo* suggests prominent roles for β-MSH and desacetyl α-MSH in energy homeostasis. *Molecular Metabolism* (2018) 17: 82–97.

Kumar R. B., Aronne L. J. Hypothalamic Inflammation: Is There Evidence for Human Obesity? *Current Obesity Reports* (2014) 3:242-247.

Lasaga M., Debeljuk L., Durand D., Teresa N. Scimonelli, Carla Caruso. Role Of-melanocytes stimulating hormone and melanocortin 4 receptor brain inflammation. *Peptides* (2008) 29:1825–1835.

Li L., Sweedler J. V. Peptides in the brain: mass spectrometry based measurement approaches and challenges. *Annual Reviews Analytical Chemistry* (2008) 1:451-483.

Li P., Bartiett M. G. A review of sample preparation methods for quantitation of small-molecule analytes in brain tissue by liquid chromatography tandem mass spectrometry (LC-MS/MS). *Analytical Methods* (2014) 6:6183-6207.

Matraszek-Zuchowska, I., Wozniak B., Posyniak A. Comparison of the Multiple Reaction Monitoring and Enhanced Product Ion Scan Modes for Confirmation of Stilbenes in Bovine Urine Samples Using LC–MS/MS QTRAP® System. *Chromatographia* (2016) 79:1003–1012.

Mokh S., Moussa F., Khoury E. E. L., Rania Nassar, Nicola Bernabò, Mohamad Al Iskandarani. Development of a new liquid chromatography-tandem mass spectrometry method for the determination of hormones in bovine muscle. *Journal of Pharmaceutical and Biomedical Analysis* (2020) 190:113550.

Mountjoy K. G., Caron A., Hubbard K., Avik Shome, Angus C. Grey, Bo Sun, Sarah Bould, Martin Middleditch, Beau Pontré, Ailsa McGregor, Paul W. R. Harris, Renata Kowalczyk, Margaret A. Brimble, Rikus Botha, Karen M. L. Tan, Sarah J. Piper, Christina Buchanan, Syann Lee, Anthony P. Coll, Joel K. Elmquist. Desacetyl-a-melanocyte stimulating hormone and a-melanocyte stimulating hormone are required to regulate energy balance. *Molecular Metabolism* (2018) 9:207-216.

Pelissari Kravchychyn A. C., Munhoz da Silveira Campos R., Landi Masquio D. C., Sofia Emanuelle de Castro Ferreira Vicente, Yasmin Alaby Martins Ferreira, Patrícia Leão Silva, Aline de Piano Ganen, Lila Missae Oyama, Lian Tock, Marco Túlio de Mello, Sérgio Tufik, Ana Raimunda Dâmaso. The Long-Term Impact of High Levels of Alpha-Melanocyte-Stimulating Hormone in Energy Balance Among Obese Adolescents. *Annals of Nutrition & Metabolism* (2018) 72:279–286.

Perroud B., Alvarado R. J., Espinal G. M., Alex R. Morado, Brett S. Phinney, Craig H. Warden. *In vivo* multiplex quantitative analysis of 3 forms of alpha melanocyte stimulating hormone in pituitary of prolyl endopeptidase deficient mice. *Molecular Brain* (2009) 2 (14):1-9.

Phetsanthad A., Vu N. Q., Yu Q., Amanda R. Buchberger, Zhengwei Chen, Caitlin Keller, Lingjun Li. Recent advances in mass spectrometry analysis of neuropeptides. *Mass Spectrometry Reviews* (2023) 42:706–750.

Plumb R. S., Gray R. D. M., Jones C. M. Use of reduced sorbent bed and disk membrane solid-phase extraction for the analysis of pharmaceutical compounds in biological fluids, with applications in the 96-well format. *Journal of Chromatography B* (1997) 694:123–133.

Ramirez C. E., Nouzova M., Michalkova V., Fernandez-Lima F., Noriega F. G. Common structural features facilitate the simultaneous identification and quantification of the five most common juvenile hormones by liquid chromatography-tandem mass spectrometry. *Insect Biochemistry and Molecular Biology* (2020) 116:103287.

Sun B., Vatanen T., Jayasinghe T. N., McKenzie E., Murphy R., O'Sullivan J. M. Desacetyl-α-MSH and α-MSH have sex specific interactions with diet to influence mouse gut morphology, metabolites and microbiota. *Scientific Reports* (2020) 10:18957.

Trivedi P., Jiang M., Tamvakopoulos C. C., Xiaolan Shen, Hong Yu, Steven Mock, Judy Fenyk-Melody, Lex H. T. Van der Ploeg, Xiao-Ming Guan. Exploring the site of anorectic action of peripherally administered synthetic melanocortin peptide MT-II in rats. *Brain Research* (2003) 977:221–230.

Zhou Y., Wong J. M., Mabrouk O. S., Robert T. Kennedy. Reducing Adsorption To Improve Recovery and *in Vivo* Detection of Neuropeptides by Microdialysis with LC-MS. *Analytical Chemistry* (2015) 87:9802-9809.

Index